Lecture Notes in Economics and Mathematical Systems

Managing Editors: M. Beckmann and W. Krelle

Econometrics

220

Masanao Aoki

Computer Science Dept
School of Engineering and Applied Science
4731 Boelter Hall
University of California
Los Angeles CALIFORNIA 90024
USA

Notes on Economic Time Series Analysis: System Theoretic Perspectives

Springer-Verlag
Berlin Heidelberg New York Tokyo 1983

Author

Prof. Masanao Aoki
The Institute of Social and Economic Research
Osaka University
6-1 Mihogaoka, Ibaraki, Osaka 567, Japan

Current Address:

4531 Boelter Hall
University of California, Los Angeles
Los Angeles, CA 90024, USA

ISBN 3-540-12696-1 Springer-Verlag Berlin Heidelberg New York Tokyo
ISBN 0-387-12696-1 Springer-Verlag New York Heidelberg Berlin Tokyo

Printing and binding: Beltz Offsetdruck, Hemsbach/Bergstr.
2142/3140-543210

To Chieko and

Thursday's children

Preface

In seminars and graduate level courses I have had several opportunities to discuss modeling and analysis of time series with economists and economic graduate students during the past several years. These experiences made me aware of a gap between what economic graduate students are taught about vector-valued time series and what is available in recent system literature.

Wishing to fill or narrow the gap that I suspect is more widely spread than my personal experiences indicate, I have written these notes to augment and reorganize materials I have given in these courses and seminars. I have endeavored to present, in as much a self-contained way as practicable, a body of results and techniques in system theory that I judge to be relevant and useful to economists interested in using time series in their research. I have essentially acted as an intermediary and interpreter of system theoretic results and perspectives in time series by filtering out non-essential details, and presenting coherent accounts of what I deem to be important but not readily available, or accessible to economists. For this reason I have excluded from the notes many results on various estimation methods or their statistical properties because they are amply discussed in many standard texts on time series or on statistics.

The notes naturally divide into three parts: Chapters 1 through 6 are preparatory to the main part of the notes. The notion of state, which is basic in representing time series by Markovian models, is introduced early in Chapter 2. Chapter 3 describes time-invariant dynamic systems, i.e., systems whose properties remain invariant with respect to shift of the origin of the time axis, which we mostly use to represent time series after suitable processings of data if necessary. Here, locations of zeros of the numerator and denominator polynomials of transfer functions are related to the notions of inverse systems, stable systems and minimum phase systems, the last appearing prominently in our later chapters. Several ways to represent time series are taken up in Chapters 4 and 5. Chapter 6 considers preliminary processings of time series data to fit economic data series into a common framework of mean zero, finite covariance weakly stationary stochastic processes. Chapters 7 through 10 constitute the main part of these notes. There I use singular value decomposition of certain matrices made up of covariances of data vectors to produce Markovian models that generate time-indexed data vectors. These models, after further refinement by maximum likelihood procedures if necessary, can be used to predict future values of the data vectors. Connection of this method with the canonical correlation method of Akaike is also explained. Chapter 11 on time series from intertemporal optimization may be of particular interest to some macroeconomists in view of recent research interests in explaining business cycles using equilibrium macroeconomic models. Identification of closed-loop systems and time series generated by dynamic models incorporating rational expectation are the final two topics of the lecture notes. Chapter 14 is the third part of the notes and contain several numerical examples mostly drawn from Japanese economic time series.

To help bridge the gap or barrier faced by someone who is not versed in the system theoretic language I have collected a number of brief but mostly self-contained accounts of the facts I use in the main body as mathematical appendices.

In preparing these notes, the author received help from many friends and colleagues. Sean Becketti of University of California, Los Angeles and Hiroshi Yoshikawa of the Institute of Social and Economic Research, Osaka University commented on an earlier draft. Leonard Silverman of Department of Electrical Engineering, University of Southern California showed me his unpublished report. Jorma Rissanen of IBM, San Jose told me of several important recent works on the time series analysis. I owe Quirino Paris of University of California, Davis a reference. Dr. Hirotsugu Akaike of the Institute of Statistical Mathematics, Tokyo made available to the author the computer programs that implement his AIC criterion. Arthur Havenner of University of California, Davis was instrumental in organizing a series of seminars at which some of the material in preliminary form was tried out. He also provided most useful comments on an earlier draft. Axel Leijohhufvud of University of California, Los Angeles helped the author by arranging for his visits to the Department of Economics at University of California, Los Angeles where a preliminary version of the notes was tried out at a graduate level economics course.

These notes were typed expertly and expeditiously by Ms. Y. Ishida, T. Kawata, K. Uto and G. Nystrom. Computations were carried out by Messrs. H. Ebara, S. Tateishi, K. Nakagawa and Ms. C. Baden.

Osaka

M.A.

TABLE OF CONTENTS

1 INTRODUCTION

Time series arise when data are collected over time, either continously or at discrete time instants, and usually on several related variables. Together they produce vector-valued, and time-indexed data which record economic activities. Economic data are often collected at regular intervals such as daily, weekly, monthly etc.

We analyze data jointly rather than singly, i.e., as vectors rather than as scalars;

(1) to uncover dynamic or structural relations among them, because some series may lead or lag other series, and there may be feedbacks between them,

and ultimately

(2) to forecast better, because modeling of a collection of time series as vector valued use related information in data jointly.

Study of time series has history much older than modern system theory. Probabilists, statisticians and econometricians all have contributed to advance our understanding of time series over the past several decades. Many well established books record their contributions. One may wonder what new results system theory can add to this well-established field and doubt if any new perspective or insight can be gained by this relative newcomer to the field. History of science shows us, however, that same problems can and have been examined with advantage by different disciplines, partly because implications of alternative assumptions are explored by researchers with different backgrounds or interests, and partly because new techniques developed elsewhere are brought in to explore areas left untouched by the discipline in which the problem originated. Although a latecomer to the field of time series analysis, system theory has brought a set of viewpoints, concepts and tools slightly different from the traditional ones, and they are effective in dealing with vector-valued time-indexed data.

We believe that system theory has provided new perspectives, and contributed new results especially on vector-valued time series that are of potential interest to economists who must deal with time series but are not experts in time series analysis and modeling. Our primary objective in writing this set of notes is to help them overcome the language barriers because these results and perspectives are stated in a language unfamiliar to them, and make these results and new tools accessible to economists in order that they may benefit from system theory in their own researches.

What are the new perspectives and results we speak of? First, how do we represent time series? Loosely put, traditional time series analysis is primarily directed toward scalar-valued data, and usually represent time series by scalar autoregressive, moving average or autoregressive-moving average models. We provide alternative modes of representing vector-valued time-indexed data, and directly connect exogenous variables at several time points with endogenous variables also at several time instants. In one mode, this connection is expressed by means of the transfer function matrices which relate the input time series with the output time series. Classical control literature also made much use of the transfer functions of dynamic systems. Modern control and system theory improves on the classical results and handle several variables simultaneously as vector-valued variables and has introduced an alternative mode of representing dynamic phenomena, called the state space or Markovian representation, by defining internal or state space variables as useful auxiliary variables. Although these two ways of representing dynamic phenomena are equivalent, they have their own advantages and disadvantages. Having two alternate ways of dealing with vector-valued time series is definitely worthwhile.

Prompted by this different viewpoint of dynamic phenomena, or different mode of descriptions thereof, and by the necessity of paying greater attention to the interrelation of vector components, system theory has introduced theoretical notions that are nontrivial only for vector-valued time series such as that of reachability, observability, and of minimal realization which are not found in the traditional i.e., scalar-value oriented time series literature. These notions turn out to be rather significant in many cosiderations on modeling of time series by minimal dimensional Markovian representations and in examining robustness of various algorithms for identification. For example, as we later show, the problem of common factors in the AR portion and MA portion of ARMA models is exactly that of minimal realization of given time series by Markovian models.

Secondly, given a time series represented in state space form how do we construct a particular model, i.e., pick dimensions and estimate parameters? How do we test for identifiability? We use canonical correlations or equivalently singular value decomposition of matrices composed of covariances known as Hankel matrices and relate Hankel matrices to construction of minimal dimensional Markovian representations of vector-valued time series, and state identifiability of closed-loop dynamics in terms of a system notion called return differences of feedback systems.

In this set of notes I try to highlight a few aspects of analysis and modeling of time series that are primarily system theoretic in origin or orientation in order to provide some new perspectives or analytical techniques. I have chosen

Hankel matrices as a unifying theme to treat time series prediction, representation of data by (lower order) state space models, and examination of identification and identifiability conditions from a common viewpoint. Singular value decomposition of certain Hankel matrices followed by suitable scaling produces so-called internally-balanced state space models of vector-valued time series. These models may further be refined if needed by additional parameter estimation steps by maximizing likelihood functions suitably modified to account for the number of parameters in the models. Two criteria due to Akaike and Rissanen are discussed.

2 THE NOTION OF STATE

Time series are regarded as being generated by as Markovian or state space dynamic systems. The concept of state is one of the key notions in dynamics. State is not a topic routinely discussed by economists or econometricians. (Harvey [1981] seems to be the only book on time series written by econometricians that mentions state space. Even he devotes only one chapter to this topic, however.) State naturally arises in problems of optimization over time. It is not an artificial mathematical construct to burden economists unnecessarily. Bellman and Dreyfus [1962] discuss many deterministic and stochastic examples illustrating this fact. See Chapter 11 as well.

Loosely put, a state vector of a deterministic dynamic system is a minimum collection of information necessary to uniquely "determine" the future evolution of the dynamic system, given future time paths of all relevant exogenous variables affecting the system including decision or choice variables. For example, in a system governed by $z_{t+1} = f(z_t, x_t)$ where x_t is an exogenous variable, the vector z_t is a state vector of this system, because z_{t+1} is uniquely determined by z_t and x_t.

Suppose dynamic equations involve some predetermined, i.e., lagged endogenous variables. An example is a system described by

$$z_{t+1} = f_t(z_t, z_{t-1}, x_t, x_{t-1}), \qquad t = 1, 2, \ldots$$

$$z_1 = f_0(z_0, x_0, z_0),$$

where x_t is the exogenous input as before.

In such a system the knowledge of z_t and the current and the future values of x, i.e., x_τ, $\tau \geq t$ is not sufficient to determine future values of z's.

The value of z_{t-1} and x_{t-1} must additionally be known before z_{t+1} can be uniquely determined. Evidently z_t alone is not a state vector of this system. Introduce additional information by defining a vector by

$$s_t = \begin{pmatrix} z_t \\ y_t \\ w_t \end{pmatrix}$$

where

$$y_t = z_{t-1} \qquad \text{and} \qquad w_t = x_{t-1}.$$

Then the relation

$$s_{t+1} = \begin{pmatrix} z_{t+1} \\ y_{t+1} \\ w_{t+1} \end{pmatrix} = \begin{pmatrix} f_t(z_t, y_t, w_t, x_t) \\ z_t \\ x_t \end{pmatrix} = F_t(s_t, x_t),$$

where $F(\cdot, \cdot)$ is defined by the above identity, shows that the knowledge of s_t plus the values of current and future exogenous variables x_τ, $\tau \geq t$ suffice to determine s_t hence z_τ, $\tau \geq t$ uniquely. Thus the vector s_t qualifies as a state vector for this dynamic system. Chapter 5 describes systematic methods for converting linear dynamic models such as ARMA models into state space form where state vectors may contain some lagged endogenous variables. The above example shows that a similar procedure works for non-linear systems as well.

When stochastic processes are involved, we must properly re-interpret the phrase uniquely "determine" in our description of the notion of state. In stochastic systems, probability laws for evolution are the best one can specify to determine uniquely future evolutions of the dynamic system in general. In special cases where probability laws can be specified by a few statistics such as first or second order moments, then they can serve as finite-dimensional state vectors. Otherwise the state vector becomes infinite-dimensional.

3 TIME-INVARIANT LINEAR DYNAMICS

Dynamic systems or their models determine the time paths of endogenous variables for given time paths of exogenous variables. If the models are represented by a (set of) differential or difference equations, then endogenous variables are obtained as the solutions of these equations using exogenous variables as the right-hand term, i.e., as the input or forcing variables, as they are called in the system literature.

Dynamics are called time-invariant or time-homogeneous if the system characteristics do not change with time. In other words, a stationary output $y(t+\tau)$ results in response to the input $u(t+\tau)$ where $u(t)$ causes the stationary output to be $y(t)$, i.e., time translation of input signals merely translates the output time function by the same amount. Time-invariance corresponds to the notion of stationarity in stochastic processes. In stochastic processes probability laws or moments are invariant with respect to translation of the processes along the time axis.

Dynamic systems are called <u>linear</u> if endogenous variables (outputs) $\alpha_1 y_1(t) + \alpha_2 y_2(t)$ correspond to exogenous variables (inputs) $\alpha_1 u_1(t) + \alpha_2 u_2(t)$ where $y_i(t)$ is endogenous variable corresponding to $u_i(t)$ alone, $i = 1, 2$. We say that the superposition principle holds for linear systems. Stability is implicitly assumed in discussing stationary outputs. Effects of nonzero initial state (initial conditions) die out with time for stable systems. Actually, this is more or less what we mean by stable dynamics. See Aoki [1976, Chapter 4] for more precise discussion of stability.

3.1 Continuous Time Systems

Because periodic signals have Fourier series representation, and because the superposition principle holds for linear dynamic systems, we know responses of a linear, time-invariant dynamic systems to any periodic input signal once we know the output to a stationary input $u(t) = e^{j\omega t}$. Denote it by $y(t)$. Then, by linearity the input $u(t+\tau) = e^{j\omega(t+\tau)} = e^{j\omega\tau} \cdot e^{j\omega t} = e^{j\omega\tau} u(t)$ produces the stationary output $e^{j\omega\tau} y(t)$. On the other hand, by the time invariance, $y(t+\tau)$ is the stationary output produced by $u(t+\tau)$. Hence assuming uniqueness of stationary outputs $y(t+\tau) = e^{j\omega\tau} y(t)$. Setting t to zero and replacing τ by t yield a relation

$$y(t) = e^{j\omega t} y(0).$$

This equation shows that the stationary output produced by the input $e^{j\omega t}$ is a constant multiple of the input. We replace the constant $y(0)$ by $H(j\omega)$ to show the explicit dependence of the constant (i.e., independent of t) on $j\omega$. This expression $H(j\omega)$, which is a complex number in general, is called the frequency response function and shows how the system responds to signals of different frequencies. For a general input $u(t)$, express it by its Fourier transform as

$$u(t) = \frac{1}{2\pi} \int_{-\infty}^{\infty} U(j\omega) e^{j\omega t} d\omega,$$

i.e., the signal is made up of periodic signal $e^{j\omega t}$ with amplitude $U(j\omega)/2\pi$. Because $e^{j\omega t}$ produces output $H(j\omega)e^{j\omega t}$, by the superposition principle the signal $u(t)$ produces the output

$$y(t) = \frac{1}{2\pi} \int_{-\infty}^{\infty} H(j\omega) U(j\omega) e^{j\omega t} d\omega.$$

Taking its Fourier transform, we can express the above as

$$Y(j\omega) = H(j\omega) U(j\omega).$$

The same relation holds when $j\omega$ is replaced by a general complex number $s = \sigma + j\omega$

(1) $$Y(s) = H(s)U(s).$$

This $H(s)$ is called the transfer function.

In the time domain a convolution of a function $h(\cdot)$ with the input $u(\cdot)$ represents the above relation

$$y(t) = \int_{-\infty}^{\infty} h(\tau)u(t-\tau)d\tau = \int_{-\infty}^{\infty} h(t-\tau)u(\tau)d\tau,$$

where $h(t)$ is obtained by the inverse Laplace transforms and is called the impulse response function.

We say a dynamic system is causal when its impulse response function vanished for the negative time argument, $h(t) = 0$ for $t < 0$. Referring to the convolution expression above, the causal system output $y(t)$ is determined by inputs $u(\tau)$, $\tau \leq t$ only, i.e., any future signal $u(t+s)$, $s \geq 0$ does not affect the value of $y(t)$ hence the system is called causal.

For continuous dynamic processes, its state space representation takes the form of the first order vector differential equation

(2) $$d\chi/dt = A\chi + Bu.$$

Its solution consists of the zero input solution (solution of the homogeneous part with $u \equiv 0$) and the zero-state solution which is the solution corresponding to zero initial condition $\chi(0) = 0$. Suppose that A is a constant matrix. Then the solution of (2) is given by

(3) $$\chi(t) = e^{At}\chi(0) + \int_0^t e^{A(t-\tau)}Bu(\tau)d\tau.$$

This can be readily verified by substitution and using the relation $de^{At}/dt = Ae^{At}$. When A is not a constant matrix, we cannot write the solution as above. Instead we have

$$\chi(t) = \phi(t, 0)\chi(0) + \int_{t0}^{t} \phi(t, \tau)Bu(\tau)d\tau,$$

where $\phi(t, \tau)$ is an $(n\times n)$ matrix called the fundamental solution matrix. It satisfies

$$d\phi(t, \tau)/dt = A\phi(t, \tau), \qquad \phi(t, t) = I.$$

In other words, $\phi(t, \tau) = e^{A(t-\tau)}$ if A is constant. Otherwise, the explicit form of $\phi(t, \tau)$ is not usually available.

With a constant A, a change of variable shows that

$$\int_0^t e^{A(t-\tau)}Bu(\tau)d\tau = \int_0^t e^{A\tau}Bu(t-\tau)d\tau.$$

We see that $e^{A\tau}B$ is the impulse response of the input τ time earlier, i.e., dynamic multiplier of $u(t-\tau)$ on $\chi(t)$. It is called impulse response because a narrow pulse (impulse)

$$u(t) = \begin{cases} 1/\varepsilon & \text{over } [t-\tau, t-\tau+\varepsilon] \\ 0 & \text{elsewhere} \end{cases}$$

approximately gives rise to $\chi(t) \simeq e^{A\tau}B$. More generally $\phi(t, \tau)$ is the impulse response function.

Linear dynamic systems whose characteristic remains the same through time are called time-invariant or time-homogeneous systems. They are more conveniently handled using the Laplace transform. The Laplace transform of $\chi(t)$ is defined by

$$X(p) = \int_0^\infty \chi(t)e^{-pt}dt$$

if the integral exists. For example, Laplace transforms are defined for all $\chi(\cdot)$ such that

$$\int_0^\infty \chi^2(t)dt < \infty.$$

Laplace transforms of impulse functions are called transfer functions.

With $\chi(0)$ zero in (3), the Laplace transform of $\chi(\cdot)$ equals

$$X(p) = \int_0^\infty dte^{-pt} \int_0^t e^{A(t-\tau)}Bu(\tau)d\tau$$

$$= \int_0^\infty e^{A\tau}Be^{-p\tau}d\tau \int_0^\infty u(v)e^{-pv}dv$$

(4) $$= H(p)U(p),$$

where U(p) is the Laplace transform of u(•) and H(p) is the Laplace transform of the impulse response function $e^{A\tau}B$. (4) is said to "transfer" the effects of "input" onto the "output", hence the name transfer functions.

We note that

$$H(p) = (pI-A)^{-1}B$$

$$= N(p)/D(p)$$

where

$$D(p) = |pI-A|$$

is the characteristic polynomial of the matrix A. See Aoki [1976, p.45] for an algorithm for calculating the numerator N(p). The transfer function is a ratio of two polynomials, a rational transfer function. The zeros of the numerator polynomial are called zeros of the transfer function. The zeros of the denominator polynomial are the poles of H(p). A rational transfer function is stable if and only if all poles lie in the left half of the complex plane. A pole in the right half plane gives rise to an exponentially growing impulse response in magnitude. For stability reasons we exclude systems with such unstable impulse responses from consideration.

3.2 Inverse Systems

What distinguishes two stable transfer functions whose zeros are mirror images of each other with respect to the imaginary axis? An example helps here. Let $H_1(p) = (p+1)/(p+2)(p+3)$ and $H_2(p) = (p-1)/(p+2)(p+3)$. Note that $H_2(p)$ differs from $H_1(p)$ by a factor $(p-1)/(p+1)$. This factor has magnitude 1 for any p on the imaginary axis because $(p-1)/(p+1) = |(p-1)/(p+1)|e^{-j\phi(\omega)}$ where $\phi(\omega) = 2\tan^{-1}\omega > 0$. Therefore $e^{j\omega t}$, when applied to $(p-1)/(p+1)$, produces $e^{j(\omega t-\phi(\omega))}$ or $e^{j\omega(t-\tau(\omega))}$ where $\tau(\omega) = \phi(\omega)/\omega$. We call $\tau(\omega)$ (phase-) delay. $H_2(p)$ has the same magnitude as $H_1(p)$. Its (phase-) delay, however, is larger than that of $H_1(p)$ by $\tau(w)$. For this reason we call transfer functions with zeros in the right half plane non-minimum phase transfer functions. The crucial point to note is that the inverse of $H_2(p)$ is unstable but $1/H_1(p)$ is stable: In the example, $1/H_2(p) = p + 6 + 12/(p-1)$ and $1/H_1(p) = p + 4 + 4/(p+1)$. The term $12/(p-1)$ produces a divergent impulse response $12e^t$ while $4/(p+1)$ produces a convergent response $4e^{-t}$. If a system has a minimum phase transfer function $H(p)$, then following it with another system, called the inverse system, with the transfer function $1/H(p)$ recovers the original impulse. Chapter 4 again takes up the inverse system. Our interest in inverse systems lies not so much in deterministic systems but rather in stochastic systems where the inverse systems are related to the notion of calculating innovation sequences (i.e., by whitening filters to convert a weakly stationary sequence with specified covariances to a white noise sequences) or in shaping filters to reproduce a given covariance sequence from a white noise sequence. We return to these topics in Chapter 10.

3.3 Discrete-Time Sequences

Economic time series are often subjected to one or more data processing procedures. These produce one or more sequences from a given one. Even when no causal or dynamic relations are implied or present, a stochastic sequence $\{y_n\}$ with a complicated sequence of covariances may be conveniently represented as the output of passing a white noise sequence, i.e., a sequence with a constant correlation coefficient, through a filter i.e., as the output of a dynamic system subject to white noises.

Input-output sequences are related by the transfer function H(z), with an argument z being L^{-1}. We can relate H(z) to the transfer function of some continuous time dynamic system by setting z to $e^{j\omega T}$ where T is a basic sampling period of the sequences, i.e., data are collected only at integer multiple of T. A stable H(z) has all its poles inside the unit circle. This fact has been discussed elsewhere. Just as transfer functions of continuous dynamic systems with the same magnitude can have different (phase) delay, by having their zeros in symmetric positions across the imaginary axis, the same variance-covariance structures can be represented by two transfer functions whose zeros are inside and outside the unit circle in the z-plane, respectively.

Suppose $y(n) = \Sigma_m h(n-m)u(m)$ where we write u(n) to indicate the value of u(t) at t = nT, with T being the period of data collection. The z-transforms of sequences are defined by

$$Y(Z) = \sum_{-\infty}^{\infty} y(n)z^{-n}, \quad H(Z) = \Sigma\, g(n)z^{-n} \text{ and } U(z) = \sum_{-\infty}^{\infty} u(n)z^{-n}.$$

The convolution relation becomes

$$Y(Z) = H(Z)U(Z).$$

This is the discrete-time version of (1). We further discuss z-transforms in Appendix A.5. The impulse response sequence $\{h_n\}$ is causal if $h_n = 0$ for negative n. For discrete-time causal dynamic systems the convolution expression becomes

$$y(t) = \sum_{n=-\infty}^{t} h(t-n)u(n) = \sum_{0}^{\infty} h(n)u(t-n).$$

A transformation $p = (1-z^{-1})/(1+z^{-1})$ translates the results for the continuous and discrete time dynamics. This is a conformal mapping of a complex plane p into another complex plane z. The imaginary axis in the p-plane becomes the unit circle $|z| = 1$. The region $Re(p) < 0$ is mapped into $|z| < 1$. In this formal way we associate two transfers functions $h_1(z)$ and $h_2(z)$ with zeros inside and outside the unti circle with a pair of transfer functions, $H_1(p)$ and $H_2(p)$ with the zeros in symmetric positions across the imaginary axis. As with the continuous transfer function, one with zero inside the unit circle produces a stable and, causal inverse. Because of the relation, $(1-z)/(1+z) = (z^{-1}-1)/(z^{-1}+1) = -(1-z^{-1})/(1+z^{-1})$, the pair z and 1/z are the mirror images in the z-plane with respect to the unit circle, corresponding to the fact that p and -p are the mirror images about the imaginary axis in the p-plane. For example, $h_1(z) = (1+0.5z^{-1})/(1+0.1z^{-1})$ produces $1/h_2(z) = (1+0.1z^{-1})/(1+0.5z^{-1})$. But $h_2(z) = (1+2z^{-1})/(1+0.1z^{-1})$ with a zero at $z = -1/0.5 = -2$ which is the mirror image of $z = -0.5$ relative to the circle $|z| = 1$, produces an unstable inverse $1/h_2(z) = (1+0.1z^{-1})/(1+2z^{-1})$.

4 TIME SERIES REPRESENTATION

Economic data are inherently noisy. We regard them as (discrete-time) stochastic processes, using the first and second order moments to characterize them. By removing known mean values from data $\{y_t\}$, the first moments can be taken to be zero. So we focus on the structure of second order moments. Thus, information contained in data is summarized by sequences of covariance matrices $\Lambda_{t,s} = E(y_t y_s')$. We attempt to duplicate the covariance matrix sequence of a given time series by solutions of a difference equation with zero-mean and finite covariance stochastic processes as inputs.

When translation of input signals along the time axis merely translates the output time functions by the same amount and leave (first and second order) moments invariant then the process is called weakly stationary. Covariance matrices of weakly stationary processes $\Lambda_{t,s}$ depend on the time difference t-s rather than on t and s separately, as Λ_{t-s}. Weakly stationary stochastic processes are modeled by time homogeneous dynamic equations. Time homogeneity and linearity alone tells us a lot about model dynamics as we have shown in Chapter 3. Otherwise time series are nonstationary and modeled by difference equations with time-varying coefficients. (Dynamics whose properties change with translation along the time axis are called time-varying or time inhomogeneous.)

To describe the future time paths of a time series we generally need the values of the current exogenous vector and values of one or more of the predetermined, i.e., lagged endogenous and exogenous variables. How many of these predetermined variables do we need? That depends on the complexity of structures generating the time series, and must eventually be estimated from data. In Chapter 3 we explained that the notion of transfer functions is natural for weakly stationary processes. A stationary output produced by the input $e^{j\omega t}$ is a constant multiple of the input

$$y(t) = H(j\omega)e^{j\omega t}.$$

This expression $H(j\omega)$, which is a complex number in general, is called the frequency response or transfer function and shows how the system responds to signals of different frequencies. The idea is that $H(j\omega)$ tells us how the effect of input is transferred onto the output. The above description is somewhat geared towards continuous-time system. However, the notion of transfer function is also natural in discrete-time processes.

Earlier we spoke of using difference equations to represent time series. In economics and econometrics literature we often find difference equations written using the lag operator L; we write $y_t - 7y_{t-1} + 4y_{t-2}$ as $(1-7L+4L^2)y_t$, for example. Generally, the equation

$$\phi(L)y_t = \psi(L)u_t, \tag{1}$$

for some (polynomial) functions $\phi(L)$ and $\psi(L)$, is one of the most common representations of the time series. When y_t and u_t are scalars, $\phi(L)$ is a polynomial (of degree p) in the lag operator L and $\psi(L)$ is another polynomial (of degree q) in L. The expressions such as $\phi(L)y_t$ are thus a convenient short-hand notation for a linear relation among $y_t, y_{t-1}, \ldots, y_{t-p}$. The mode of representation associated with (1) is basically a reduced form in economics. The ratio $\psi(L)/\phi(L)$ is the transfer function. When y and/or u are vectors, ϕ and ψ generally become matrices where each element is a polynomial in L and $\phi^{-1}\psi$ is the transfer function matrix. We use scalar valued time series to describe various models, and return later to vector-valued models. When the polynomial ψ is 1, we call the model autoregressive (AR). A model with $\phi = 1$ is called a moving average (MA) model. A 'generic' model is a combination of these two, where neither ϕ nor ψ is one. So the model is called autoregressive-moving average (ARMA). When u_t contains non-random exogenous component, we sometimes speak of ARMAX models.

A systematic or deterministic component of a time series is usually removed before further processing of data. One common way to remove a deterministic component is to difference data one or more times. When y_t is not the original data series but is a processed one by taking the difference of the original time series, we called the original model autoregressive-integrated moving average (ARIMA) model.

Older books on economic dynamics such as Allen [1966], Baumol [1970] and Gandolfo [1971] use this approach. They typically work their way up from models described by first order differential (or difference) equations, to models governed by the second order dynamics and finally to models of higher order dynamics. Dynamics are also introduced into $\{y_t\}$ when we carry out some data processing operations on them. The effects of such processing on y_t can also be conveniently expressed using lag polynomials or transfer functions in L in general. See Appendices A.1 and A.5 for a concise exposition of difference equations and z-transforms.

The system literature favors the other approach, and uses the state space or Markovian representation, which expresses dynamics by a first order difference equation for an internal, or an auxiliary set of variables z_t called the state vector

$$z_{t+1} = Az_t + Bu_t \quad : \quad \text{state equation,}$$

and relate y_t and u_t to the state vector by

$$y_t = Cz_t + Du_t \quad : \quad \text{output or data (observation) equation.}$$

This representation is closer in spirit to structural forms than reduced forms in economics and econometrics. We can easily establish the equivalence of these two representations.

An example may clarify the difference in representation. Suppose a stochastic process is described by

$$\phi(L)y_t = u_t$$

where

$$\phi(L) = 1 - \alpha_1 L - \alpha_2 L^2 - \dots - \alpha_p L^p.$$

Then the transfer function h(L) connects u_t to y_t by

$$y_t = h(L)u_t$$

where

$$h(L) = 1/\phi(L).$$

The same time series represented by the state space (or Markovian) model is

$$z_{t+1} = Az_t + bu_t$$

where

$$z_t' = (y_{t-p+1}, y_{t-p+2}, \dots, y_t),$$

$$A = \begin{pmatrix} 0 & 1 & 0 & \dots\dots & 0 \\ 0 & 0 & 1 & 0 \dots & 0 \\ & & & & \\ 0 & \dots\dots\dots & & 0 & 1 \\ \alpha_p & \dots\dots\dots\dots & & & \alpha_1 \end{pmatrix}, \quad b = \begin{pmatrix} 0 \\ \vdots \\ 0 \\ 1 \\ \alpha_1 \end{pmatrix},$$

$$y_t = (0 \dots 0\ 1)z_t + u_t.$$

We return to discuss general "conversion" methods for the two modes of representations in the next chapter.

One may wonder about the wisdom of this second, 'obviously' round-about way of describing time series. However, ease of expressing solutions of vector-valued first order difference equations definitely make this a worthwhile mode of representing time series. We later elaborate on this point in detail. State space representation of dynamic systems has several features to recommend it over the other, even though the latter is more familiar to economists and econometricians. One is the standardization of model representation achieved by this procedure. Models are always stated as the first order difference equation for the state vector. Only the dimension of the state vectors vary from time series to time series. Although only practice and experience really brings home the superior nature of this representation for some purposes, it should be apparent that economy of thought is achieved, and that uniform representation facilitates development of solution algorithms.

Models in either of the two modes of time series representation determine the time paths of the endogenous variables, given time paths of exogenous variables, i.e., the sequence $\{y_t\}$ is obtained as the solutions of appropriate difference equations with the exogenous variables as inputs. We now turn to the nonuniqueness of such representations. More than one difference equation driven by the same input stochastic processes yield the same covariance sequences as the data $\{y_t\}$. A related question is this: When do we construct an AR model $\phi(L)y_t = u_t$, and when an MA model $y_t = \psi(L)u_t$? Or, is there any restriction on writing $\psi(L)$ as $1/\phi(L)$?

Adequate answers to these questions depend on the locations of the zeros of the numerator polynomials of the transfer functions as we next demonstrate. The problem is related to the notion of the inverse system of Chapter 3. We call a dynamic system with the representation $y_t = h(L)u_t$, the inverse of another

$u_t = h(L)y_t$, because the roles of inputs or the forcing terms are reversed in these two dynamic systems. To be useful, these two dynamics must both be stable. This means that zeros of both the numerator polynomial and denominator polynomial must be stable. A simple example illustrates: Let two positive constants c_0 and c_1 define two sequences by

$$\eta_t = c_0\varepsilon_t - c_1\varepsilon_{t-1},$$

and

$$\eta_t = c_1\varepsilon_t - c_0\varepsilon_{t-1},$$

where $0 < c_1 < c_0$, and $\{\varepsilon_t\}$ is a mean-zero white noise sequence such that

$$E\varepsilon_t = 0, \qquad E\varepsilon_t\varepsilon_s = \sigma^2\delta_{t,s},$$

where $\delta_{t,s}$ is one for $t = s$ and zero otherwise. These two sequences have identical variances and covariances, var ζ_t = var $\eta_t = (c_0^2+c_1^2)\sigma^2$ and cov (ζ_t, ζ_{t-1}) = cov $(\eta_t, \eta_{t-1}) = c_0c_1$. All other covariances are zero. Yet only one is causally invertible:

$$\zeta_t = (c_0-c_1z^{-1})\varepsilon_t$$

or

$$\varepsilon_t = \frac{1}{c_0-c_1z^{-1}}\zeta_t = \frac{1}{c_0}\{1+az^{-1}+a^2z^{-2}+ \dots\}\zeta_t,$$

since $a = -c_1/c_0$ has magnitude less than 1. The same point can be made by a slightly different sequence $\eta_t = \varepsilon_t + \rho\varepsilon_{t-1}$, $|\rho| > 1$ is not causally invertible while $\zeta_t = \varepsilon_t + \rho\varepsilon_{t-1}$, $|\rho| < 1$ is.

These two sequences become distinguishable when we examine their phase characteristics: Let T = 1 for simplicity. Define the transfer function

$$H(e^{j\omega}) = c_0 - c_1e^{-j\omega} = \sqrt{c_0^2 + 2c_0c_1\cos\omega + c_1^2}\; e^{j\tau(\omega)}$$

where

$$\tan \tau(\omega) = c_1 \sin \omega / (c_0 + c_1 \cos \omega)$$

and

$$H_y(e^{j\omega}) = c_1 - c_0 e^{-j\omega} = \sqrt{c_0^2 + 2c_0c_1\cos \omega + c_1^2}\; e^{j\tilde{\tau}(\omega)}$$

where

$$\tilde{\tau}(\omega) = c_0 \sin w / (c_1 + c_0 \cos \omega).$$

Clearly $\tilde{\tau}(\omega) > \tau(\omega)$, i.e., the transfer function for $\{\eta_t\}$ is larger than that for $\{\zeta_t\}$. For continuous dynamics, a transfer function which is rational in the complex variable s is stable if and only if all zeros of the denominator lie in the left half of the s complex plane. A zero of the denominator in the right half plane gives rise to an impulse response exponentially growing in magnitude. For stability reasons we exclude systems with such unstable impulse responses from consideration.

5 EQUIVALENCE OF ARMA AND STATE SPACE MODELS

This section explains how non-Markovian models can be converted into Markovian models in simple and systematic ways. This conversion uses changes of variables that may at first sight appear to be arbitrary, but is in fact quite natural once the procedures are understood.

Although quite simple, and possible in many different ways, the conversion procedures of this section incorporate some thoughts and care to achieve "minimal dimensional" Markovian models. Other seemingly simpler procedures may obtain state space models with too much redundant information or irrelevant information. Generally speaking, non-minimal dimensional models are to be avoided because such model representations may effectively prevent efficient optimization calculations because of needless high dimensions. If informational redundancy is the only problem, non-minimal dimensional models are not to be frowned upon too much. However, they may suffer from other technical deficiencies not obvious at first. For example, algorithms for filtering, estimation and control are conventionally stated for minimal dimensional models and may break down or require special handling if applied to nonminimal dimensional systems because technical conditions (such as positive-definiteness of matrices assumed in the algorithms) may be violated.

We discuss only scalar systems because the question of minimal dimensional representation does not occur and enables us to concentrate on the conversion procedures. Our method generalizes to vector valued systems in a natural way. See Aoki [1981; Appendix A] for an example. We caution the reader that some other seemingly straightforward extension of the conversion procedures sometimes lead to inconvenient state space models.

5.1 AR Models

Time series models can be put into state space form in several ways. It is probably most convenient to start with AR models.

We use the letter L to denote a backward shift, i.e., $Ly_t = y_{t-1}$. Then an AR model $y_t - \alpha_1 y_{t-1} - \alpha_2 y_{t-2} - \dots - \alpha_p y_{t-p} = u_{t-1}$ can be put as

(1) $$\phi(L) y_t = u_{t-1}$$

where

$$\phi(L) = 1 - \alpha_1 L - \alpha_2 L^2 - \dots - \alpha_p L^p.$$

Here y's and u's are taken to be scalar-valued. Introduce p auxiliary variables by setting

$$\begin{aligned} \chi_1(t) &= y_{t-p+1}, \\ \chi_2(t) &= y_{t-p+2}, \\ &\vdots \\ \chi_{p-1}(t) &= y_{t-1}, \\ \chi_p(t) &= y_t. \end{aligned}$$

Advancing time by one, and by the above definitions we can write

$$\begin{aligned} \chi_1(t+1) &= \chi_2(t), \\ \chi_2(t+1) &= \chi_3(t), \\ &\vdots \\ \chi_p(t+1) &= y_{t+1} \\ &= \alpha_1 y_t + \dots + \alpha_p y_{t-p+1} + u_t \\ &= \alpha_1 \chi_p(t) + \dots + \alpha_p \chi_1(t) + u_t. \end{aligned}$$

Define a (state) vector of these variables as $\chi(t) = (\chi_1(t), \dots, \chi_p(t))'$. Then $\chi(t)$ evolves with time according to a first order difference equation, called the state transition equation

(2) $\chi(t+1) = A\chi(t) + bu_t$,

where

$$A = \begin{pmatrix} 0 & 1 & 0 & \cdots & \cdots & 0 \\ 0 & 0 & 1 & \cdots & \cdots & 0 \\ 0 & \cdots & \cdots & \cdots & 0 & 1 \\ \alpha_p & \cdots & \cdots & \cdots & \cdots & \alpha_1 \end{pmatrix}, \qquad b = \begin{pmatrix} 0 \\ \vdots \\ 0 \\ 0 \\ 1 \end{pmatrix},$$

and y_t is related to $\chi(t)$ by an algebric relation, called an observation or output equation

(3) $y_t = (0 \ldots 0 \quad 1)\chi(t)$.

The pair of equations (2) and (3) constitutes a state space representation of (1).

If the right hand side of (1) is u_t rather than u_{t-1}, then redefine $\chi_p(t)$ to be $y_t - u_t$. The definitions for the other components of the state vector χ remain the same. Equation (2) remains valid by redefining b to be b' = $(0 \ldots 0 \quad 1 \quad \alpha_1)$. Equation (3) is replaced by $y_t = (0 \quad 0 \quad \ldots \quad 1)\chi(t) + u_t$.

5.2 MA Models

Let

$$y_t = \psi(L)\varepsilon_t$$

where

$$\psi(L) = \beta_0 + \beta_1 L + \ldots + \beta_q L^q.$$

Introduce state vector components by

$$\chi_{1t} = \varepsilon_{t-1},$$

$$\chi_{2t} = \varepsilon_{t-2},$$

$$\vdots$$

$$\chi_{qt} = \varepsilon_{t-q}.$$

Then advancing t by one we note that

$$\chi_{1t+1} = \varepsilon_t,$$
$$\chi_{2t+1} = \chi_{1t},$$
$$\vdots$$
$$\chi_{q,t+1} = \chi_{q-1,t}.$$

Hence the state vector

$$\chi_t = \begin{pmatrix} \chi_{1t} \\ \vdots \\ \chi_{qt} \end{pmatrix}$$

evolves with time according to

$$\chi_{t+1} = \begin{pmatrix} 0 & \dots & \dots & 0 \\ 1 & 0 & \dots & 0 \\ \dots & \dots & \dots & \dots \\ 0 & \dots & 1 & 0 \end{pmatrix} \chi_t + \begin{pmatrix} 1 \\ 0 \\ \dots \\ 0 \end{pmatrix} \varepsilon_t,$$

and y_t is related to the state vector by the equation

$$y_t = (b_1, \dots, b_q)\chi_t + b_0\varepsilon_t.$$

5.3 ARMA Models*

Next consider a time series model described by

(4) $$y_t - \alpha_1 y_{t-1} - \dots - \alpha_p y_{t-p} = \beta_0 u_t + \beta_1 u_{t-1} + \dots + \beta_{p-1} u_{t-p+1}.$$

It can be put into a form analogous to (1),

$$\phi(L)y_t = \psi(L)u_t,$$

where $\phi(L)$ is as defined above and we define

$$\psi(L) = \beta_0 + \beta_1 L + \dots + \beta_{p-1} L^{p-1}.$$

Some of the β's may be zero. The state space representation of (4) can be obtained in two steps: First define v_t by

---

* See next page.

---

* In converting ARMA models, non-minimal dimensional models may arise. Several examples of this are found in Chow [1975]. In one such example he (p.153) turns a vector ARMA model

$$y_t = A_{1t}y_{t-1} + \dots A_{mt}y_{t-m} + C_{0t}x_t + \dots + C_{nt}x_{t-n} + u_t$$

into a state space form

$$\chi_t = A_t\chi_{t-1} + B_t x_t + Du_t$$

by introducing a vector

$$\chi_t = \begin{pmatrix} y_t \\ \vdots \\ y_{t-m+1} \\ x_t \\ \vdots \\ x_{t-n+1} \end{pmatrix} \qquad B_t = \begin{pmatrix} C_{0t} \\ 0 \\ \vdots \\ 0 \\ I \\ 0 \\ \vdots \\ 0 \end{pmatrix}$$

and appropriate matrices A_t, and D.

Clearly, χ_t is a state vector because its knowledge and a future time path of exogenous disturbances u's and a control time path x's eneables us to specify uniquely (the probability distributions) of the future values of χ's. Notice, however, a peculiar feature of his dynamic matrix; A_t contains a zero row submatrix, corresponding to the identity matrix of the B matrix, i.e., the matrix A_t is of the form

$$A_t = \left(\begin{array}{ccc:ccc} x & \dots & x & x & \dots & x \\ x & & & & 0 & \\ & & x & & & \\ 0 & \dots & & 0 & \dots & 0 \\ & & & x & & \\ & 0 & & & & \\ & & & & & x \end{array}\right)$$

where x marks nonzero submatrices. The state vector is not controllable in the sense we discuss later (see Aoki [1976; Chapter 3], for example), even though a subvector made up of y_t and its lagged values is controllable. Although nothing is theoretically wrong with this representation because the relevant part of χ_t is controllable it may be inconvenient to store redundant information in a computer, and standard algorithms for minimizing quadratic costs subject to linear dynamics may not apply without modifications.

(5) $\phi(L)v_t = u_t.$

Then (4) can be written as

(6) $y_t = \psi(L)v_t.$

The sequence $\{v_t\}$ generated by (5) has the same form as (1), hence can be put into state space form

$$\chi(t+1) = A\chi(t) + bu_t, \tag{7}$$
and
$$v_t = (0 \ \ldots \ 0 \quad 1)\chi(t) + u_t,$$

by introducing the vector $\chi(t) = (\chi(t), \ldots, \chi_p(t))'$ where $\chi_1(t) = v_{t-p+1}$, $\ldots, \chi_{p-1}(t) = v_{t-1}$ and $\chi_p(t) = v_t - u_t$. The state vector $\chi(t)$ is next related to y_t of (6) by

$$\begin{aligned} y_t &= \beta_0 v_t + \beta_1 v_{t-1} + \ldots + \beta_{p-1} v_{t-p+1} \\ &= \beta_0(\chi_p(t) + u_t) + \beta_1\chi_{p-1}(t) + \ldots + \beta_{p-1}\chi_1(t) \\ &= c'\chi(t) + \beta_0 u_t \end{aligned} \tag{8}$$

where

$$c' = (\beta_{p-1}, \ldots, \beta_1, \beta_0).$$

Collecting (7) and (8) together, a state space representation of the ARMA model (4) becomes

$$\begin{aligned} \chi(t+1) &= A\chi(t) + bu_t, \\ y_t &= c'\chi(t) + \beta_0 u_t. \end{aligned} \tag{9}$$

There are other ways of putting ARMA models into state space form. We discuss two such representations; a controllable and an observable representation.

For example, given

$$y_t = \alpha_1 y_{t-1} + \ldots + \alpha_p y_{t-p} + \beta_0 u_t + \ldots + \beta_p u_{t-p},$$

let

$$\chi_1(t) = y_t - \beta_0 u_t.$$

Then we can write the above as $\chi_1(t) = L(\alpha_1 y_t + \beta_1 u_t) + \ldots + L^p(\alpha_p y_t + \beta_p u_t)$,

or

$$L^{-1}\chi_1(t) = \chi_1(t+1) = \alpha_1 y_t + \beta_1 u_t + \chi_2(t),$$

where we introduce a new variable by

$$\chi_2(t) = L(\alpha_2 y_t + \beta_2 u_t) + \ldots + L^{p-1}(\alpha_p y_t + \beta_p u_t).$$

Eliminating y_t by the first equation, the above becomes

$$\chi_1(t+1) = \alpha_1\chi_1(t) + (\alpha_1\beta_0 + \beta_1)u_t + \chi_2(t).$$

Rewrite the definitional equation for $\chi_2(t)$ as

$$L^{-1}\chi_2(t) = \chi_2(t+1) = \alpha_2 y_t + \beta_2 u_t + \chi_3(t)$$
$$= \alpha_2\chi_1(t) + (\alpha_2\beta_0 + \beta_2)u_t + \chi_3(t),$$

where we introduce a new variable by

$$\chi_3(t) = L(\alpha_3 y_t + \beta_3 u_t) + \ldots + L^{p-2}(\alpha_p y_t + \beta_p u_t).$$

Continue in this way until we reach

$$\chi_p(t) = L(\alpha_p y_t + \beta_p u_t)$$

or

$$\chi_p(t+1) = \alpha_p\chi_1(t) + (\alpha_p\beta_0 + \beta_p)u_t.$$

Collecting $\chi_1(t+1), \ldots, \chi_p(t+1)$ as a column vector, we define it as the state vector. Then it obeys the state transition equation

$$\chi(t+1) = \begin{pmatrix} \chi_1(t+1) \\ \vdots \\ \chi_p(t+1) \end{pmatrix} = A\chi(t) + u(t), \tag{9'}$$

and the output equation becomes $y_t = (1 \quad 0 \quad \ldots \ldots \ldots \quad 0)\chi(t) + \beta_0 u_t$

where

$$A = \begin{pmatrix} \alpha_1 & 1 & 0 & \ldots & \ldots & \ldots & 0 \\ \alpha_2 & 0 & 1 & 0 & \ldots & \ldots & 0 \\ \alpha_{p-1} & \ldots & \ldots & \ldots & \ldots & \ldots & 1 \\ \alpha_p & 0 & \ldots & \ldots & \ldots & \ldots & 0 \end{pmatrix} \quad \text{and} \quad b = \begin{pmatrix} \alpha_1\beta_0+\beta_1 \\ \vdots \\ \alpha_p\beta_0+\beta_p \end{pmatrix}.$$

This representation is called an observable canonical form.

Compare the alternative state space representation of (9) and (9'). The representation (9) has a simple b vector with a complicated c vector, while (9') has a complicated b vector and a simple c vector. The form (9) is called a controllable representation because the p-vectors b, Ab, ..., $A^{p-1}b$ are easily seen to be linearly independent because they have the structure

$$b = \begin{pmatrix} 0 \\ \cdot \\ \cdot \\ \cdot \\ 0 \\ 1 \end{pmatrix}, \qquad Ab = \begin{pmatrix} 0 \\ \cdot \\ 0 \\ 0 \\ 1 \\ x \end{pmatrix}, \qquad A^2b = \begin{pmatrix} 0 \\ \cdot \\ 0 \\ 1 \\ x \\ x \end{pmatrix}, \qquad A^{p-1}b = \begin{pmatrix} 1 \\ x \\ \cdot \\ \cdot \\ \cdot \\ x \end{pmatrix},$$

Examples Second-order systems are used to give an example of observable canonical form and its dual or controllable canonical form. The precise sense in which these canonical forms are duals of each other is also indicated. Consider

$$y_t = -\alpha_1 y_{t-1} - \alpha_2 y_{t-2} + \beta_0 u_t + \beta_1 u_{t-1} + \beta_2 u_{t-2}.$$

Write this as a nested sequence of lag operations:

$$y_t - \beta_0 u_t = L\{(-\alpha_1 y_t + \beta_1 u_t) + L(-\alpha_2 y_t + \beta_2 u_t)\}.$$

Let

$$\chi_1(t) = y_t - \beta_0 u_t.$$

Substitute y_t out by $\chi_1(t) + \beta_0 u_t$ to yield

$$L^{-1}\chi_1(t) = -\alpha_1 \chi_1(t) + \gamma_1 u_t + L(-\alpha_2 y + \beta_2 u_t)$$

where

$$\gamma_1 = \beta_1 - \alpha_1 \beta_0.$$

Letting $\chi_2(t)$ equal to $L(-\alpha_2 y + \beta_2 u_t)$, and recalling that $L^{-1}\chi_1(t) = \chi_1(t+1)$ we derive

$$\chi_1(t+1) = -\alpha_1\chi_1(t) + \gamma_1 + u_t + \chi_2(t)$$

and

$$\chi_2(t+1) = -\alpha_2 y + \beta_2 u_t$$
$$= -\alpha_2\chi_1(t) + \gamma_2 u_t$$

where

$$\gamma_2 = \beta_2 - \alpha_2\beta_0.$$

Collect the above two terms to form

(10) $$\begin{pmatrix}\chi_1(t+1)\\ \chi_2(t+1)\end{pmatrix} = \begin{pmatrix}-\alpha_1 & 1\\ -\alpha_2 & 0\end{pmatrix}\begin{pmatrix}\chi_1(t)\\ \chi_2(t)\end{pmatrix} + \begin{pmatrix}\gamma_1\\ \gamma_2\end{pmatrix}u_t,$$

and

$$y_t = (1\quad 0)\begin{pmatrix}\chi_1(t)\\ \chi_2(t)\end{pmatrix} + \beta_0 u_t.$$

This is an observable canonical form.

The same system can be written as a sequence in L^{-1}:

$$L^{-2}(y_t - \beta_0 u_t) + L^{-1}(\alpha_1 y_t - \beta_1 u_t) + (\alpha_2 y_t - \beta_2 u_t) = 0$$

or

$$(L^{-2} + \alpha_1 L^{-1} + \alpha_2)(y_t - \beta_0 u_t) = (\gamma_1 L^{-1} + \gamma_2)u_t,$$

or

$$(L^{-2} + \alpha_1 L^{-1} + \alpha_2)(\gamma_2 L + \gamma_1)\chi_t = (\gamma_1 L^{-1} + \gamma_2)u_t$$

where we define χ_t by

$$y_0 - \beta_0 u_t = (\gamma_2 L + \gamma_1)\chi_t.$$

Thus

$$y_t = \beta_0 u_t + (\gamma_1 + \gamma_2 L)\chi_t.$$

Rewriting the above as

$$(\gamma_1 L^{-1} + \gamma_2)\{(L^{-1} + \alpha_1 + \alpha_2 L)\chi_t - u_t\} = 0,$$

let

$$L^{-1}\chi_t = -(\alpha_1 + \alpha_2 L)\chi_t + u_t.$$

Redefine χ_t as χ_{1t} and let $L\chi_{1t} = \chi_{2t}$.

Then

$$y_t = (\gamma_1, \gamma_2)\begin{pmatrix}\chi_{1t}\\ \chi_{2t}\end{pmatrix} + \beta_0 u_t$$

and

$$\chi_{2t+1} = \chi_{1t},$$

$$\chi_{1t+1} = -\alpha_1\chi_{1t} - \alpha_2\chi_{2t} + u_t,$$

or

(11) $$\begin{pmatrix}\chi_{1t+1}\\ \chi_{2t+1}\end{pmatrix} = \begin{pmatrix}-\alpha_1 & -\alpha_2\\ 1 & 0\end{pmatrix}\begin{pmatrix}\chi_{1t}\\ \chi_{2t}\end{pmatrix} + \begin{pmatrix}0\\ 1\end{pmatrix}u_t,$$

and

$$y_t = (\gamma_1, \gamma_2)\begin{pmatrix}\chi_{1t}\\ \chi_{2t}\end{pmatrix} + \beta_0 u_t.$$

This relation is not surprising when we remember that optimal predictions involve minimization of the estimation error or orthogonal projection. We return to this topic later.

This is an example of the controllable canonical form. Comparing (10) and (11), we note that the dynamic matrices are transposes of each other, and the roles of b and c are interchanged. We call a dynamic system

$$\chi_{t+1} = A\chi_t + bu_t$$

$$y_t = c'\chi_t + du_t$$

a dual of another dynamic system

$$\chi_{t+1} = A'\chi_t + cu_t$$

$$y_t = b'\chi_t + du_t.$$

This correspondence $A' \leftrightarrow A$, and $b \leftrightarrow c$ is the precise sense in which these two systems are dual. Intertemporal minimization of a quadratic cost subject to a linear dynamics and the optimal one-step ahead predictor (Kalman filter) are the duals in this sense. See Appendix A.14 also.

6 DECOMPOSITION OF DATA INTO CYCLICAL AND GROWTH COMPONENTS

6.1 Reference Paths and Variational Dynamic Models

Time series analysis often assumes zero-mean processes. To accommodate this assumption, non mean-zero components, especially secular growth of time series must be removed before we can begin. More often, economic time series are decomposed into three components; seasonal components, secular growth components and cyclical components or fluctuations about the secular growth paths. Structural information is then extracted from the remaining cyclical components or fluctuations about the growth paths to help predict future fluctuations or to discern patterns of cyclical co-movements of elements making up the time series, or better to characterize business cycles, and so forth.

This section describes a systematic way for decomposing time series into the reference path component and fluctuations about them when well specified models producing time series are posited.

This approach provides one benchmark analysis of time series. The other benchmark approach does not presuppose any such well-specified models for secular growth paths but merely extracts smoothly varying paths subject or some statistical regularity conditions.

Either method of decomposition will leave mean-zero, finite-variance and weakly stationary time series as our object to study. If such time series are given to us to begin with, we can of course dispense with this preliminary phase of data processing and directly construct Markovian (or state-space) models or their equivalent ARMA models. This aspect is discussed later.

How to decompose a given time series theory is a theoretical and numerical question sometimes addressed by time series analysts. See Akaike [1980] or Shiskin and Plewes [1978] on seasonal adjustment for example. After seasonal variations are removed, economic time series are then decomposed into secular growth components and cyclical fluctuations. This decomposition can of course be done simultaneously. Some time series are published in already seasonally adjusted forms.

We shall use the word "reference" instead of (secular) growth to denote non-zero mean components, and speak of decomposing time series into the reference paths and fluctuations or variations about the reference paths. What constitutes reference paths largely depends on the amount of structural or theoretical information we possess or wish to bring to models that are producing the time series in question.

In an extreme instance, a balanced growth path of a neoclassical macroeconomic model with constant or variable growth rates may be used as the reference path.* In cases where exogenous variables are present, then their conditionally expected values may be used together with the hypothesized (macro)economic models to define a set of reference time paths for endogenous variables of the models. In these cases fluctuations about the reference paths can be described by variational models derived from the original models that define or produce the reference time paths. This view is important and useful because detrended log-linear models often used in econometrics can be justified precisely this way. We return to this point and develop it further shortly.

The other extreme attitude adopted by some practitioners eschews any such theoretical construct of the economy as unjustified and unwarranted in

---

* Aoki [1980] is one such example.

view of existing economic theory. The only distinguishing features, then, between the growth and cyclical components are the relative frequencies involved. Growth components, for example, are "known" to vary much more 'smoothly' than cyclical components. Operationally, this distinction may be made by an (arbitrary) assumption on behavior of some higher order difference (such as the second order) of growth paths being randomly varying with variances comparable with those for the cyclical components. Hodrick and Prescott [1981] take such an approach, for example.

6.2 Log-linear Models as Variational Models

We now show that if a model is specified that produces a reference time path, then the model for fluctuations around it is the same as its detrended, log-linear model. (Such a model is called the variational model in the systems literature.) See Aoki [1981; Chapter 2] for additional details on the variational models. We first establish this connection of the variational models producing cyclical or fluctuating movements about the reference time paths with familiar log-linear models.

Log-linear economic models arise in at least two ways; one is familiar to economists while the other seems less so.

An example will illustrate the difference. Consider a money demand function specified in product form as

$$M/P = e^{-\alpha i} Y^{\beta}. \tag{+}$$

Define $m = \ell nM$, $p = \ell nP$ and $y = \ell nY$. Take the logarithm of (+) to obtain

$$m - p = -\alpha i + \beta y.$$

This is indeed a familiar demand for real balances specified to be linear in logarithms of variables, except for the interest rate i.

If the variables are measured as deviations from some reference (long-run equilibrium, for example), we can redefine the variables:

$$m = \ell n(M/M_0),\ p = \ell n(P/P_0) \text{ and } y = \ell n(Y/Y_0)$$

where the subscript "0" denotes reference. Now, because $\ell nM - \ell nM_0 = \ell n(M/M_0)$ etc., the lower case variables measure deviations from the reference. Note, however, that this approach works only for functional relations which are specified in product form.

This approach can not handle relations such as $Y = H + I$ even when the variables are measured as deviations from a reference, i.e., $y = \ell n(Y/Y_0) = \ell n\frac{H + I}{(H + I)_0}$.

Now the second approach (which is the one in Chapter 2 of Aoki [1981]) is used: Suppose $Y = H + I$ and define lower case letters by

$$Y = Y_0(1 + y), \qquad H = H_0(1 + h), \qquad I = I_0(1 + r).$$

Then

$$Y_0(1 + y) = H_0(1 + h) + I_0(1 + r)$$

or because $Y_0 = H_0 + I_0$, we can write this equation as

$$Y_0y = H_0h + I_0r$$

or

(*) $$y = (H_0/Y_0)h + (I_0/Y)r.$$

By definition, $\ell n(y/y_0) = \ell n(1 + y) \simeq y$, $\ell n(H/H_0) = \ell n(1 + h) \simeq h$, and $\ell n(I/I_0) = \ell n(1 + r) \simeq r$, i.e., (*) is the log-linear model resulting from the variational approach, and the lower case variables measure deviations of the logarithms of the corresponding upper case letters.

This way of deriving log-linear relations from the variational approach is thus more general because it does not require any specific functional form.

Another example may make this point clear. One way to specify a consumer price index of a small open economy is to posit

$$P_I = P^{\alpha}(EP^*/P)^{1-\alpha}$$

where α is the fraction of domestic goods in the consumption bundle. Its log-linear version is

(#) $$p_I = \alpha p + (1 - \alpha)(e + p^* - p).$$

It is <u>not</u> necessary, however, to adopt such a specific functional form to produce log-linear relation. Just specify an arbitrary functional relation with usual sign restrictions

(0) $$P_I = F(\overset{+}{P}, E\overset{+}{P}{}^*/P).$$

Its reference value is

$$(P_I)_0 = F(P_0, (EP^*/P)_0).$$

Now define

$$P_I = (P_I)_0(1 + p_I),\ P = P_0(1 + p),\ P^* = P_0^*(1 + p^*) \text{ and } E = E_0(1 + e).$$

Then retaining only the first order terms in the Taylor expansion of (0) as the lower case variables, we deduce that the variational variables are related by

$$p_I = ap + b(P^* + e - p)$$

where a and b are elasticities evaluated on the reference paths and are given as $a = (\partial F/\partial P)_0/(P_0/F_0)$ and $b = \{\frac{1}{F}\frac{\partial F}{\partial(EP^*/P)}\frac{EP^*}{P}\}_0$.

The variational model of a growth model will produce a secular growth or trend time path. Many examples are worked out in Aoki [1981].

7 PREDICTION OF TIME SERIES

7.1 Prediction Space

The covariance matrix of a stacked data vector $(y_1', y_2', \dots y_N')'$ of a mean-zero weakly stationary process $\{y_t\}$ has a special structure: A submatrix $\Lambda_0 = Ey_1y_1'$ is located along the main diagonal, the matrix $\Lambda_\ell = Ey_{\ell+1}y_1'$ along the ℓ-th diagonal below the main diagonal, and $\Lambda_{-\ell} = Ey_1y_{\ell+1}' = \Lambda_\ell'$ along the ℓ-th diagonal above the main diagonal. This covariance matrix is a block Toeplitz matrix because the same submatrices are arranged in the same way that elements are arranged in Toeplitz matrices.

When we relate stacked predicted future vectors $y_{t+s|t}$, $s = 1, \dots$ to the current and past exogenous noise vectors, we obtain another matrix of special structure (not unrelated to Toeplitz matrices as discussed in Appendix A.13), called a Hankel matrix. More importantly perhaps, Hankel matrices naturally arise in our attempt to predict the future realization of $\{y_t\}$ and in constructing its Markovian model by calculating the covariance matrix between a stacked data vector $(y_t', y_{t-1}', \dots, y_{t-N}')'$ and stacked future realizations $(y_{t+1}', \dots, y_{t+K}')'$ for some positive N and K. Hankel matrices also arise in several other contexts as well; in calculating dynamic multipliers or impulse responses, in approximating impulse responses by those of some low-order dynamics, and in some identification conditions.

Use of an impulse response sequence or an MA form is one way to state the dynamic response of a discrete-time system disturbed by exogenous impulses (shocks). Let a (matrix) sequence $\{H_i\}$ relates y_t to current and past shocks by

(1) $$y_t = H_0\varepsilon_t + H_1\varepsilon_{t-1} + H_2\varepsilon_{t-2} + \dots$$

where y_t is the current observation and ε's are the exogenous shocks.

Suppose that a Markovian model of $\{y_t\}$ is given by

$$\chi_{t+1} = A\chi_t + B\varepsilon_t, \qquad (2)$$
$$y_t = C\chi_t + D\varepsilon_t.$$

The z-transform of (2) easily relates y's to ε's which is the z-transform counterpart of (1)

$$y(z) = H(z)U(z)$$

where

$$U(z) = \Sigma\ \varepsilon_k z^{-h}, \quad y(z) = \Sigma\ y_h z^{-h},$$

and

$$H(z) = \Sigma\ H_i z^{-i}$$
$$= C(zI-A)^{-1}B + D.$$

We recognize the last to be the transfer function of the dynamics (2).

The impulse response (matrix) H_i is the dynamic multiplier (matrix) which measures the effect of a past action or disturbance (vector) ε_{t-i} on the current data (vector) y_t. Appendix A.18 further discusses the multipliers. From (3) we can state H_i in the system parameters:

$$H_i = \begin{cases} CA^{i-1}B, & i > 1, \\ D, & i = 0. \end{cases} \qquad (3)$$

The H's are called Markov parameters. If $\{\varepsilon_t\}$ is a weakly stationary mean-zero white noise sequence, i.e., $E\varepsilon_t = 0$, $E\varepsilon_t\varepsilon_s' = \delta_{ts}I$, then H_i can also be interpreted as the covariance between y_t and the disturbance i periods earlier

$$H_i = E(y_t\varepsilon_{t-i}').$$

In a multivariate ARMA model, the exogenous and endogenous variables are related by

$$\phi(L)y_t = \psi(L)\varepsilon_t$$

where $\phi(L)$ and $\psi(L)$ are matrices of polynomials, i.e., $\phi(L) = \Sigma_0^p A_i L^i$ and $\psi(L) =$

$\Sigma_0^q B_i L^i$, where A_i and B_i are matrices of appropriate dimensions. In other words, each element of $\phi(L)$ and $\psi(L)$ are polynomials in L. Formally inverting $\phi(L)$, y_t and ε_t are related by

$$y_t = H(L)\varepsilon_t$$

where

$$H(L) = \phi(L)^{-1}\psi(L)$$
$$= \sum_0^{\infty} H_i L^i$$

is the transfer function. (It is called the left matrix fraction description (MFD) of the transfer function.) The transfer function is causal if $H_i = 0$ for negative i and $\|H(0)\|$ is finite.

Now, advance time in (1) successively to write y_{t+1}, y_{t+2}, ... and obtain their conditional expectation (orthogonal projection onto the subspace spanned by ε_t, ε_{t-1}, ...) expressed as a linear combination of the current and past ε's. In the notes we use the notation $y_{t+i|t}$ to denote the conditional mean of y_{t+i} given the information available at time t, i.e., ε_t, ε_{t-1}, ..., i.e.,

$$y_{t+i|t} = E(y_{t+i}|\varepsilon_t,\varepsilon_{t-1}, \ldots) = H_i\varepsilon_t + H_{i+1}\varepsilon_{t-1} + \ldots, \; i = 1, 2, \ldots .$$

When these predictions of future observations are stacked into an infinite dimensional column vector, this vector is related to the stacked conditioning vectors by the (infinite) matrix $\mathbb{H}$,

$$\text{(4)} \qquad \begin{pmatrix} y_{t+1|t} \\ y_{t+2|t} \\ \cdot \\ \cdot \\ \cdot \end{pmatrix} = \mathbb{H} \begin{pmatrix} \varepsilon_t \\ \varepsilon_{t-1} \\ \cdot \\ \cdot \\ \cdot \end{pmatrix}$$

where

$$\mathbb{H} = \begin{pmatrix} H_1 & H_2 & H_3 & \cdots \\ H_2 & H_3 & H_4 & \cdots \\ H_3 & & & \\ \cdot & & & \\ \cdot & & & \\ \cdot & & & \\ \cdot & & & \end{pmatrix}.$$

This matrix has the same submatrix H_i along counter diagonal lines (lines running from lower left to upper right, perpendicular to diagonal lines). A matrix with this feature is called a Hankel matrix. So we call $\mathbb{H}$ a block Hankel matrix. We later use another block Hankel matrix in which the submatrices H_i are not the impulse response matrices but are covariance matrices $Ey_{t+\ell}y_t'$, $\ell = 1, 2, \ldots$.

When we stack only a finite number of predictions $y_{t+s|t}$, $s = 1, 2, \ldots, N$, then we note that an upper left-hand corner of $\mathbb{H}$

$$\mathbb{H}_N = \begin{pmatrix} H_1 & \cdots & H_N \\ \cdot & & \cdot \\ \cdot & & \cdot \\ \cdot & & \cdot \\ H_N & & H_{2N-1} \end{pmatrix}$$

relates $y_{t+1|t}$ through $y_{t+N|t}$ to ε_t through ε_{t-N+1}. If y_t is p-dimensional then H_N is an (Np x Np) matrix.

From the definitional relations in (4) we observe that $\mathbb{H}$ is a product of two semi-infinite matrices

$$\mathbb{O} = \begin{pmatrix} C \\ CA \\ CA^2 \\ \cdot \\ \cdot \\ \cdot \end{pmatrix} \quad \text{and} \quad \mathbb{C} = [B, AB, \ldots, A^{N-1}B \ \ldots].$$

The (N x N) matrix $\mathbb{H}_N$ is the product of $\mathbb{O}_N$ and $\mathbb{C}_N$, two finite submatrices of $\mathbb{O}$ and $\mathbb{C}$:

$$\mathcal{H}_N = \mathcal{O}_N \mathcal{C}_N . \tag{5}$$

These two matrices are exactly the observability and controllability matrices respectively so important in system theory bacause they appear as important technical conditions in many optimization and filtering problems. See Aoki [1976] for details on these matrices.

Now, if (2) is a correct minimal-dimensional state space representation of $\{y_t\}$, then the rank of both $\mathcal{O}_N$ and $\mathcal{C}_N$ are equal to the dimension of the state vector χ_t. Let $n = \dim\chi_t$. We know from system theory that the ranks of $\mathcal{O}_N$ and $\mathcal{C}_N$ are at most n. From this fact and the relation $\mathcal{H}_N = \mathcal{O}_N \mathcal{C}_N$, we conclude that rank $\mathcal{H}_N$ is at most n. The rank is exactly n once $Np \geq n$, for minimal dimensional state space models, i.e., if (2) is controllable and observable. The regular pattern of submatrices in $\mathcal{H}$ tells us that the (row) rank of $\mathcal{H}$ can not be larger than that of $\mathcal{H}_N$ for some suitable N. The row rank of $\mathcal{H}_N$ therefore tells us the dimension of a state vector which can represent $\{y_t\}$ via a state space model (2). The state vector dimension need not be an integer multiple of the dimension of y. Just because a component of y_t, say the third component, enters the state vector does not mean that the third component of y_{t-1} also enters into the state vector. We later discuss in detail choices of basis vectors to span the row space of $\mathcal{H}$, i.e., to span the predictor space.

When we calculate covariance between a stacked data vector $(y_t', y_{t-1}', \ldots, y_{t-N+1}')'$ and the future vector $(y_{t+1}', \ldots, y_{t+N}')$, we obtain an important example of the Hankel matrix

$$E\begin{pmatrix} y_{t+1} \\ \cdot \\ \cdot \\ \cdot \\ \cdot \\ y_{t+N} \end{pmatrix} [y_t', y_{t-1}' \ldots y_{t-N+1}'] = \begin{pmatrix} \Lambda_1 & \Lambda_2 & \cdots & \Lambda_N \\ \Lambda_2 & \cdots\cdots & & \Lambda_{N+1} \\ \cdot & & & \\ \cdot & & & \\ \Lambda_N & \cdots\cdots & & \Lambda_{2N-1} \end{pmatrix} . \tag{6}$$

This matrix is important for us because we construct a state-space model of the time series $\{y_t\}$ by operating on this and related Hankel matrices. To relate this Hankel matrix to the one we just discussed, suppose that $\{y_t\}$ is modeled by

$$(7)\qquad \begin{cases} \chi_{t+1} = A\chi_t + F\varepsilon_t \\ y_t = C\chi_t + \varepsilon_t \end{cases}$$

where $\{\varepsilon_t\}$ is the usual mean-zero, serially uncorrelated exogenous process. (We later obtain a model of this type by constructing a Kalman filter in Chapter 10.)

Then the covariance submatrix Λ_t is equal to $Ey_t y_0'$ by the weak stationarity of the y_t process. On the assumption that χ_0 is uncorrelated with ε_0, hence with all ε_t, $t > 0$ as well, and by solving (7) $y_t = \varepsilon_t + C(A^t\chi_0 + \Sigma_{\tau=0}^{t-1} A^{t-1-\tau}F\varepsilon_\tau)$ the covariance has the structure*

$$(8)\qquad \Lambda_t = Ey_t y_0' = CA^{t-1}M,$$

where

$$M = A\Pi C' + F$$

and

$$\Pi = E\chi_0\chi_0' = E\chi_t\chi_t'.$$

The last equality follows if A is a stable matrix because $\{\chi_t\}$ will then be weakly stationary.

Comparing (8) with (4), we note that the Hankel matrix of (6) can also be factored as shown by (5) because Λ's and H's have the same structure. The matrix B is simply replaced with M in $\mathfrak{C}_N$, i.e., when $\mathfrak{C}_N$ of (5) is replaced with $(M, AM, \ldots, A^{N-1}M)$. Before we turn to the important topic of estimating A, C and M, we note an invariance property: The rank of $\mathbb{H}$ is invariant with respect to a similarity transformation, i.e., an equivalent choice of another state vector does not alter the rank of $\mathbb{H}$. This is easy to verify.

---

* Under some technical conditions, sample covariance matrices $(1/K)\Sigma_{t=0}^{K-1} y_{t+\ell}y_t'$ converge to $E(y_{t+\ell}y_t')$ as $K \to \infty$.

7.2 Equivalence

Suppose we are presented with two Markovian models

$$(S)\qquad \begin{cases} z_{t+1} = Az_t + Bx_t, \\ y_t = Cz_t + Dx_t, \end{cases}$$

and

$$(S^*)\qquad \begin{cases} w_{t+1} = Fw_t + Gx_t, \\ y_t = Hw_t + Dx_t. \end{cases}$$

In addition, we are told that the state vectors z_t and w_t are related by a nonsingular transformation T

$$z_t = Tw_t.$$

Then these two models are different representations of the same dynamics with respect to two different coordinate systems if the matrices satisfy the relations

$$F = T^{-1}AT,$$

$$G = T^{-1}B,$$

and

$$H = CT.$$

When the transfer matrices of these two models are examined, they are equal because

$$D + H(zI - F)^{-1}G = D + CT(zI - T^{-1}AT)^{-1}T^{-1}B$$
$$= D + C(zI - A)^{-1}B.$$

Alternatively, from (4) we can say that these two equivalent systems possess the same set of Markov parameters

$$HF^iG = (CT)(T^{-1}A^iT)(T^{-1}B) = CA^iB,\ i = 0,\ 1,\ \ldots$$

We call (S) and (S*) equivalent if such a nonsingular transformation T exists and write S ~ S* using "~" to denote equivalence. Hence the entries of the Hankel matrices are the same for equivalent model representations.

7.3 Cholesky Decomposition and Innovations

Wishing to predict future realizations of a time series is one of the basic reasons for analyzing time series. The inverse system we introduced in Chapter 3 is linked to a special and easy case of prediction which we mention in passing. We give a more detailed exposition later in several related chapters.

Suppose $\{y_t\}$ is a scalar-valued weakly stationary process. Predicting future y's from the data set, $y_1, y_2, \ldots, y_N$, becomes particularly simple if we can express y's in an MA form

$$y_t = \psi(L)e_t, \text{ where } \psi(L) = 1 + \beta_1 L + \ldots + \beta_q L^q,$$

and $\{e_t\}$ is a serially uncorrelated mean zero weakly stationary process with varience σ^2. This is because $E(y_t | e_{t-1}, \ldots e_1)$ is given by $\beta_1 e_{t-1} + \beta_2 e_{t-2} + \ldots + \beta_q e_{t-q}$, if we know $e_1, e_2, \ldots, e_{t-1}$. How do we obtain these e's? They can be obtained by factoring the covariance matrix of the data vector $(y_1, y_2, \ldots, y_{t-1})'$. Later we show that e's can also be generated by Kalman filters.

Let Z be the covariance matrix of a stacked vector $z = (y_1, \ldots, y_N)'$, i.e., $Z = E(zz')$. Factoring Z into the product form $\sigma^2 CC'$ where the matrix C is a lower triangular matrix with ones along the main diagonal, we can represent y's as a linear combination of e's,

$$z = Cu$$

where $u = (e_1, \ldots, e_N)'$, because $E(zz') = CE(uu')C' = \sigma^2 CC' = Z$.

This factorization of the covariance matrix expresses y's as a linear combination of uncorrelated disturbances. Conversely, the uncorrelated shocks can be constructed from y's by inverting the matrix C

$$u = C^{-1}z$$

or

$$e_j = \Sigma_{k=1}^{j} a_{jk} y_k$$

where the matrix $A = (a_{jk})$ is the inverse of the matrix C. Note that this is in an AR form. We can check that e's thus constructed are serially uncorrelated.

This factorization, called the Cholesky decomposition, is exactly the same as the Gram-Schmidt orthogonalization procedure frequently used in nonlinear programming algorithms. In the geometry of the mean-zero random variables with finite covariances, this analogy becomes exact. The Cholesky factorization orthogonalizes the data vectors into uncorrelated noise vectors. Let us rephrase this fact using the notion of "innovations" because we frequently deal with this notion in later sections, especially in discussing Kalman filters and estimating model parameters from data. A set of N independent (mean-zero finite variance) vectors $\tilde{y}_i$, $i = 1, \ldots, N$ is called innovations of the set of (mean-zero finite variance) data vectors, y_i, $i = 1, \ldots, N$, if for any k, the σ-field generated by $\tilde{y}_i$, $i = 1, \ldots, k$ is identical to that generated by y_i, $i = 1, \ldots, k$. In the geometric language of the Hilbert space, the subspaces spanned by $\tilde{y}_i$, $i = 1, \ldots, k$ and y_i, $i = 1, \ldots, k$ coincide. Intuitively, a set of innovation vectors carries the same amount of information as that contained in a set of data vectors.

Such innovation vectors don't always exist for general data vectors. For Gaussian random vectors, however, the innovation vectors always exist. We construct them by the Gram-Schmidt, or if you prefer, the Cholesky decomposition method:

$$\tilde{y}_1 = y_1$$

$$\tilde{y}_i = y_i - E(y_i | y_1 \ldots y_{i-1}), \quad i = 2, \ldots, N.$$

For Gaussian vectors, we know that $\tilde{y}_i$ is independent of $y_1 \ldots y_{i-1}$. We also know that the conditional expectations of random vectors that are jointly Gaussian are linear in the conditioning vectors. We thus write the above as $\tilde{y}_i = y_i - \Sigma_{j=1}^{i-1} a_{ij} y_j$, or

$$\begin{pmatrix} \tilde{y}_1 \\ \tilde{y}_2 \\ \cdot \\ \cdot \\ \cdot \\ \tilde{y}_N \end{pmatrix} = \begin{pmatrix} I & & & \\ -a_{21} & I & & 0 \\ \cdot & & & \\ \cdot & & & \\ \cdot & & & \\ -a_{N1} & \cdot & \cdot & \cdot & -a_{NN-1}I \end{pmatrix} \begin{pmatrix} y_1 \\ \cdot \\ \cdot \\ \cdot \\ \cdot \\ \cdot \\ y_N \end{pmatrix}.$$

Because this block lower triangular matrix is nonsingular, the data vectors are expressible as linear combinations of the innovation vectors. For $i > j$, we calculate $E(\tilde{y}_i\tilde{y}_j') = E[E(\tilde{y}_i\tilde{y}_j'|y_1 \ldots y_{i-1})]$ to see that it is zero. Here $\tilde{y}_j$ is measurable with respect to the σ-field generated by $y_1 \ldots y_{i-1}$, i.e., $\tilde{y}_j$ lies on the subspace spanned by $y_1 \ldots y_{i-1}$ by construction. Hence $E(\tilde{y}_i\tilde{y}_j'|y_1 \ldots y_{i-1})$ $= E(\tilde{y}_i|y_1 \ldots y_{i-1}) \cdot \tilde{y}_j' = E(\tilde{y}_i)\tilde{y}_j' = 0$. Here, we use the independence of $\tilde{y}_i$ and $y_1 \ldots y_{i-1}$. Similarly for the case where $i < j$. For Gaussian vectors, uncorrelatedness is equivalent to independence. Thus, $\tilde{y}_i$, $i = 1, \ldots, N$, are innovations as was to be proved.*

As we noted above, the prediction is easy with this form. We have the representation

$$y_{t+1} = e_{t+1} + c_{t+1,t}e_t + \ldots + c_{t+1,1}e_1.$$

Denoting the conditional mean of y_{t+1}, given $e_t, \ldots, e_1$ by $y_{t+1|t}$ we can write

$$y_{t+1|t} = c_{t+1,t}e_t + \ldots + c_{t+1,1}e_1.$$

For weakly stationary processes the coefficients should depend only on the time differences, i.e., $c_{t+1,s}$ becomes c_{t+1-s}. We return to this and other points later.

---

* The Cholesky decomposition, however, suffers from computational problems. It is a linearly convergent algorithm and does not converge fast near the solution, see Pagano [1976]. Quadratically convergent algorithms for solving the algebraic Riccati equations are reported in the systems literature. They can be applied to provide efficient factorization algorithms for the spectrum. We return to this topic elsewhere. See Hewer [1971] or Molinari [1975] also.

8 SPECTRUM AND COVARIANCES

8.1 Covariance and Spectrum

The discrete Fourier transform of a finite data of a real mean zero time series $\{x_n\}_{n=0}^{N-1}$, with a regular sampling interval T, is defined by

$$\hat{X}(\omega) = \sum_{0}^{N-1} x_n e^{-j\omega nT}$$

where NT is the total time span covered by data points $x_0, x_1, \ldots, x_{N-1}$. It is the same as the truncated z-transform when we let $z = e^{j\omega T}$. Because x_n is a random variable, so is $\hat{X}(\omega)$.

Its covariance is calculated to be

$$\text{(1)} \qquad \begin{aligned} E(\hat{X}(\omega)\hat{X}(\omega)') &= \sum_n \sum_m R(n-m)e^{-j\omega(n-m)T} \\ &= \sum_{u=-(N-1)}^{N-1} (N-|u|)R(u)e^{-j\omega uT}, \end{aligned}$$

where

$$Ex_n x_m' = R(n-m).$$

Note that $Ex_m x_n' = R(m-n) = R'(n-m)$. Divide (1) by N and let N goes to infinity. Define the limit $S(\omega)$ as

$$\begin{aligned} S(\omega) &= \lim_{N\to\infty} \frac{1}{N} E(\hat{X}(\omega)\hat{X}(\omega)') \\ &= \int_{-\infty}^{\infty} R(u)e^{-j\omega uT}. \end{aligned}$$

This is called the (power) spectrum or the spectral density of the time series $\{x_n\}$.

This equation is of the form of a Fourier series expansion, hence by the inversion formula for the Fourier transform we recover R(k) by

$$R(k) = \frac{1}{2\pi}\int_{-\pi}^{\pi} S(\omega)e^{j\omega kT}d\omega.$$

Of particular interest is the covariance of x_0

$$Ex_0x_0' = R(0) = \frac{1}{2\pi}\int_{-\pi}^{\pi} S(\omega)d\omega.$$

The z-transform of the covariance sequence (which is also called the covariance generating function) is defined by

$$S(z) = \sum_{-\infty}^{\infty} R(n) z^{-n}.$$

We recognize that the spectral density is obtained by setting $z = e^{j\omega T}$ in the above. The assumption that $\Sigma\, R(k)^2 < \infty$ ensures the $S(e^{j\omega T})$ is well-defined in mean-square sense.*

From the relation $R(-n) = R'(n)$, we note that the z-transform of the covariances or the covariance generating function satisfy a relation

$$\begin{aligned} S(z^{-1}) &= \sum_{-\infty}^{\infty} R(n) z^{n} \\ &= \sum_{-\infty}^{\infty} R'(-n) z^{n} \\ &= S'(z). \end{aligned}$$

Alternatively, a covariance generating function can be written in a sum form

$$S(z) = G(z) + G(z^{-1})$$

where Re $G(e^{j\omega T}) > 0$. The function $S(z)$ is positive on $|z| = 1$. When R's are scalar, $G(e^{j\omega T})$ is even in ω:

$$\begin{aligned} S(e^{j\omega T}) &= R(0) + \sum_{1}^{\infty} R(n) e^{-j\omega nT} + \sum_{-1}^{-\infty} R(n) e^{-j\omega nT} \\ &= R(0) + \sum_{1}^{\infty} R(n) e^{-j\omega nT} + \sum_{1}^{\infty} R(-n) e^{j\omega nT} \\ &= R(0) + 2 \sum_{1}^{\infty} R(n) \cos \omega nT. \end{aligned}$$

To summarize, a spectrum $S(z)$ is the z-transform of a covariance sequence. Theoretically it satisfies the next three properties:

---

* The literature often uses the covariance generating function defined by

$$S(x) = \Sigma_{-\infty}^{\infty} R(n) x^{n}.$$

Here x is merely a place marker with no instrinsic meaning. The z-transform is the generating function when x is identified with z^{-1}.

(i) $S'(1/z) = S(z)$ (sometimes called parahermintian)

(ii) Analytic on $|z| = 1$,

and

(iii) $S(z) > 0$ on $|z| = 1$.

The sum form of S(z) shows that (i) is true. Functions satisfying (ii) and (iii) are called positive real functions.

Now regard a mean-zero stationary stochastic process y(t) as being outputs of a time-invariant linear causal dynamic system with another mean-zero stationary stochastic process u(t) as inputs:

$$y_t = \sum_{n=0}^{\infty} h_n x(t-n)$$

where $\{h_n\}$ is the impulse response sequence of this dynamics. By causality h_n is zero for all negative n. The discrete transfer function H(z) is given as the one-sided z-transform

$$H(z) = \sum_{0}^{\infty} h_n z^{-n}$$

or

$$H(e^{j\omega T}) = \sum_{0}^{\infty} h_n e^{-j\omega nT}.$$

Easy calculations show that the output covariance matrix is given by

(2)
$$\begin{aligned} R_{yy}(k) &= Ey(t+k)y'(t) \\ &= E[\{\sum_{m=0}^{\infty} h(m)x(t+k-m)\}\{\sum_{\ell=0}^{\infty} h(\ell)x(t-\ell)\}]' \\ &= \sum_{m}\sum_{\ell} h(m)R_x(k+\ell-m)h'(\ell), \end{aligned}$$

where $R_x(n)$ is the input covariance matrix, and $R_{yy}(k)$ that of output. By definition the spectrum of the y series equals

(3)
$$S_{yy}(\omega) = \sum_{k=-\infty}^{\infty} R_{yy}(k)e^{-j\omega kT},$$

and that of the x series is given by

$$S_x(\omega) = \sum_{n=-\infty}^{\infty} R_x(n)e^{-j\omega nT}. \tag{4}$$

Substitute (2) into (3) and use (4) to rewrite the spectral density of y_t in terms of that of x_t as

$$\begin{aligned} S_{yy}(\omega) &= \sum_{\tau=-\infty}^{\infty} \sum_{m=0}^{\infty} \sum_{\ell=0}^{\infty} h(m)R_x(\tau+\ell-m)h'(\ell)e^{-j\omega\tau T} \\ &= \sum_{m=0}^{\infty} h(m)e^{-j\omega mT} \sum_{\tau=-\infty}^{\infty} R_x(\tau+\ell-m)e^{-j\omega(\tau+\ell-m)T} \sum_{\ell=0}^{\infty} h'(\ell)e^{j\omega\ell T} \\ &= H(e^{j\omega T})S_x(\omega)H'(e^{-j\omega T}). \end{aligned} \tag{5}$$

This important equation relates the spectral densities of the output and input via the transfer function. It is a form of spectral factorization results. Let the variable z correspond with $e^{j\omega T}$. Then we can factor the spectrum thus

$$S_{yy}(z) = H(z)S_x(z)H(z^{-1})^*, \tag{6}$$

where * denotes conjugate transpose.

A serially uncorrelated input sequence is called a white noise sequence, i.e., $R_\ell = EX_{n+\ell}x'_n = 0$ for $\ell \neq 0$. Then the spectrum $S_x(z)$ is a constant independent of z. The spectral density of a dynamic system with a white noise sequence as input can be factored, then, as $H(z)\Sigma H(z^{-1})$ where Σ is the noise covariance matrix, $Ex_n x'_n = \Sigma$.

8.2 Spectral Factorization

The previous section calculates the spectrum of $\{y_t\}$, given its model dynamics or its transfer function, and a mean-zero white noise sequence as its input. The spectral factorization can be thought to be the converse process of generating $\{y_t\}$ as the output of a linear dynamic system driven by white noise, given the spectrum, or equivalently the covariance generating function of the y-process.

From the previous section we know that the covariance or the correlation coefficient of a real-valued process $\{y_t\}$ is real and even, and that the covariance generating function is made up of the sum of a function $G(\cdot)$ evaluated at z as well as at z^{-1}, $S(z) = G(z) + G(z^{-1})$, hence $S(z) = S(z^{-1})$, and that $S(z) > 0$ on $|z| = 1$ becuase the Toeplitz matrix $[R_{|i-j|}]$ is positive semi-definite for any m, $1 \leq i, j \leq m$. Because the coefficients are real, zeros of S(z) are either real or occur in complex conjugate pairs in the complex z-domain. In addition, because S(z) equals $S(z^{-1})$, if $z = z_1$, is a zero, so is z_1^{-1}. The zeros of S(z) hence occur in fours, with the possible exception of zeros that are exactly on the unit circle $|z| = 1$. The latter occurs in twos (complex conjugate pairs) unless $z = \pm 1$. By collecting appropriate factors, then, we can factor S(z) in a form corresponding to (6)

$$S(z) = W(z)W'(z^{-1})$$

where W(z) collects all zero lying in $|z| \leq 1$. The zero on $|z| = 1$ are equally allocated to W(z) and $W'(z^{-1})$. This is the scalar version of the Spectral Factorization theorem. A basic result for vector-valued process is:

<u>Theorem</u> Let S(z) be a real rational full rank covariance generating function. Then it can be factored as $S(z) = W(z)\Sigma W^*(z)$ where W(z) is real, rational, stable, of minimum phase, and $\Sigma' = \Sigma > 0$.

Consequently, $W^{-1}(z)$ is analytic in $|z| > 1$. We can then write for $|z| > 1$, $W^{-1}(z) = \Sigma_0^\infty c_k z^{-k}$, where the Taylor series exapnsion is valid in $|z| > 1$. The matrix $W^{-1}(z)$ is the z-transform of a stable causal (one-sided) dynamics, hence $W^{-1}(z)$ is a causally stable dynamic system called whitening filter, and $\varepsilon_t =$

$W^{-1}(z)y_t$ is the input white noise. We have already mentioned another way to factor a spectrum by Cholesky factorization of the covariance matrix. A third way, to be discussed subsequently, is to generate innovation sequences by Kalman filters and factors spectrum accordingly.

Let

$$S(z) = \sum_{-m}^{m} R_h z^{-h}.$$

Then

$$z^m S(z) = \sum_{2m}^{m} R_{m+r} z^{-r}$$
$$= \sum_{0}^{2m} R_{r-s} z^s: \text{ polynominal of degree } 2m.$$

A stochastic process taking its value in a real Euclidean space has a real-valued spectrum $S(e^{j\omega})$, i.e, $S(z)^* = S'(z^{-1})$. So if z_k is a zero of $S(z)$, then so is z_k^{-1}, if z_k is real. If z_k is complex, then z_k^* is also a zero.

Let $\{\gamma_h, \gamma_h^*\}$ be a set of complex roots of $m^m S(z) = 0$ with $|\gamma_h| > 1$ and also containing half of those roots $|\gamma_h| = 1$. Let ρ_j be a real root. Then we can factor $z^m S(z)$ as

$$z^m S(z) = \text{const}\left(\prod_{k=1}^{h} (z - \gamma_k)(z - \gamma_k^*) \prod_{j=1}^{\ell} (z - \gamma_j)\right)\left(\prod_{k=1}^{k} (z - \gamma_k^{-1})(z - \gamma_k^{*-1}) \prod_{j=1}^{\ell} (z - \rho_j^{-1})\right)$$

where

$$2k + \ell = m.$$

Let $z\beta(z) = \pi_{k=1}^{h}(z - \gamma_k)(z - \gamma_k^*)\pi_{j=1}^{\ell}(z - \rho_j)$. Then noting that $z^{-1} -\gamma = 1/z - \gamma = 1-\gamma z/z = \gamma^{-1}-z/z$. or $z[z^{-1} - \gamma] = -(z - \gamma^{-1}) = -z(z^{-1} - \gamma)$, we see that $\beta(z)$ has no zero in $|z| < 1$. If there is no γ_n or ρ_j of modulus 1, then $\beta(z)$ has no zero in $|z| \leq 1$, i.e., is a z-transform of a strictly minimum delay i.e., minimum phase filter.

System theoretic construction provides an alternative to direct spectral factorization of covariance sequences. The covariance generating functions $S(z)$ are naturally given as a sum $S(z) = G(z) + G'(z^{-1})$ because of $R_h = R'_{-h}$. Its spectral factorization expresses it as $S(z) = W(z)W^*(z^{-1})$ where $W(z)$ is analytic in $|z| \geq 1$ and of minimum phase, i.e., has zero inside the unit circle, and rank $W(z) = r$ in $|z| \geq 1$ if rank $S(z) = r$. The matrix $W(z)$, called the spectral factor, is unique up to left multiplication by an orthogonal, real-valued constant matrix. The function $G(z)$ is called positive real in the systems literature if it is analytic in $|z| \geq 1$, $G(z) + G'(z) \geq 0$ and $G(\infty)$ is finite.

We now describe an algorithm for calculating the spectral factor due to Anderson et al., [1974]. One of the system theoretic results on positive realness is that $G(z)$ is positive real if there exists a symmetric positive semi-definite matrix P such that

(7) $$M(P) = \begin{pmatrix} APA'-P, & APC'+\Gamma \\ (APC'+\Gamma)', & CPC'+2I \end{pmatrix} \geq 0,$$

where (A, Γ, C) is a minimal realization of $G(z)$, i.e.,

$$G(z) = I + C(zI - A)^{-1} ,$$

where rank $(\Gamma, A\Gamma, \ldots,) = r = \text{rank}(C', A'C', \ldots,)$.

This is easily established. Suppose such a P exists. Then factor $M(P)$ as

(8) $$M(P) = \begin{pmatrix} \Gamma \\ I \end{pmatrix} \Sigma [\Gamma' \quad I]$$

and construct

(9) $$W(z) = I + C(zI - A)^{-1} .$$

We can show that $W(z)\Sigma W'(z^{-1}) = G(z) + G'(z^{-1})$ by straightforward substitution when $\Gamma\Sigma$, $\Sigma\Gamma'$ and $\Gamma\Sigma\Gamma'$ are substituted out by the corresponding expressions from

(7).

To recapitulate: Let $S(z)$ be a rational spectrum with full rank for almost all z. It is given by $\Sigma_{-\infty}^{\infty}\Lambda_\ell z^{-\ell}$ where $\Lambda_0 = D$ and $\Lambda_\ell = CA^{\ell-1}M$, $\ell > 1$. Then the spectral factrization theorem tells us that $S(z)$ can be uniquely factored as $W(z)\Sigma W'(z^{-1})$, $\Sigma = \Sigma' > 0$ where $W(z)$ has all poles and all zeros inside the unit disc, $|z| < 1$, i.e., the poles of $W^{-1}(z)$ are also all inside the unit disc and $\lim_{z\to\infty} W(z) = I$. The spectrum $S(z)$ can also be written as $S(z) = G(z) + G'(z^{-1})$. By construction of $S(z)$, the matrix $G(z)$ is readily given by $D/2 + C(zI - A)^{-1}M$ which is realized as the transfer function of the innovation model

$$z_{t+1} = Az_t + M\varepsilon_t,$$
$$y_t = Cz_t + \frac{D}{2}\varepsilon_t.$$

A spectral factor $W(z)$ of $S(z)$ is realizable as the transfer function of a dynamic model

$$z_{t+1} = Az_t + \Gamma\varepsilon_t$$
$$y_t = Cz_t + \varepsilon_t,$$

i.e.,

$$W(z) = I + C(zI - A)^{-1}\Gamma,$$

where

$$\Gamma = K\Sigma^{-1},$$
$$\mathrm{cov}\,\varepsilon_t = \Sigma,$$
$$M = APC' + K,$$
$$D = CPC' + \Sigma,$$
$$P = APA' + K\Sigma K',$$

and

$$\Lambda_\ell = Ey_\ell y_0' = CA^{\ell-1}M,$$
$$D = Ey_0y_0'.$$

We return to these topics in Chapter 10.

Spectral factorization naturally arises in filtering problems and in control problems. Consider a Markovian model

$$\chi_{t+1} = A\chi_t + u_t,$$
$$y_t = C\chi_t + v_t,$$

where $\{u_t, v_t\}$ are jointly serially uncorrelated zero-mean processes with covariance

$$\mathrm{cov}\begin{pmatrix} u_t \\ v_t \end{pmatrix} = \begin{pmatrix} Q & N \\ N' & R \end{pmatrix}_t .$$

Its (discrete) spectrum matrix is

$$S(z) = R + C(zI - A)^{-1}N + N'(z^{-1}I - A')C' + C(zI - A)^{-1}Q(z^{-1}I - A')^{-1}C'.$$

This can be factored in terms of a matrix called the return difference matrix

$$T(z) = I + C(zI - A)^{-1}K,$$

i.e.,

$$S(z) = T(z)(R + CPC')T'(z^{-1}),$$

where

$$K = (APC' + N)(R + CPC')^{-1}$$

is the optimal Kalman filter gain and where P is the positive definite solution of the algebraic Riccati equation

$$P = APA' - (APC' + N)(R + CPC')^{-1}(CPA' + N') + Q.$$

This has been shown by several people. See Shaked [1979] or chapter 10 for example.

Its dual problem is the optimal regulator problem: minimize J where

$$J = \Sigma_0^{\infty}(\chi_{t+1}'Q\chi_{t+1} + u_t'Ru_t).$$

Subject to

$$\chi_{t+1} = A\chi_t + Bu_t.$$

Here the optimal feedback signal is

$$u_t = -K\chi_t$$

with

$$K = R^{-1}B'P.$$

The matrix P is the solution of the same Riccati equation. Here the discrete return difference matrix is given by

$$T(z) = I + K(zI - A)^{-1}B$$

and

$$T'(z^{-1})(R + B'PB)T(z) = R + B'(z^{-1}I-A')^{-1}Q(zI - A)^{-1}B = \psi(z)$$

is the form of the factorization.

Optimal regulator problems and the optimal filtering problems are called dual because the expressions for the regulator gains and filtering gains obey the same equations under suitable one-to-one correspondence. See Appendix A.14 for further detail.

8.3 Computational Aspects

Sample Covariance Matrices

In the earlier section we listed three theoretical properties of spectral densities. When an expression for theoretical spectrum is approximated by replacing true covariances with sample covariances the approximate spectrum may or may not satisfy all the three properties.

Sample covariances of $\{y\}$ are commonly calculated from a finite data $y_0, \ldots, y_{N-1}$ by

$$\hat{R}_k = \frac{1}{N} \sum_{i=0}^{N-k-1} y_{i+k}y_i', \qquad k = 0, \ldots, N-1. \tag{10}$$

This estimate is consistent but is biased. However, this approximation leads to an approximate spectrum which satisfies positive-semi-definiteness. The next example due to van Zee [1981] shows that the unbiased estimates obtained by replacing 1/N by 1/(N - k) may lead to approximate spectrum which is indefinite. For this reason, $\hat{R}_k$ of (8) is preferred.

Example Let N = 3 and $(y_0, y_1, y_2) = (1, 0, -1)$. With (N - k) replacing N, sample covariances are

$$\hat{R}_0 = \frac{1}{3}(1 + 1) = 2/3,$$
$$\hat{R}_1 = 0,$$
$$\hat{R}_2 = -1.$$

However, this approximation to the Hankel matrix

$$\mathcal{H}_3 = \begin{pmatrix} 2/3 & 0 & -1 \\ 0 & 2/3 & 0 \\ -1 & 0 & 2/3 \end{pmatrix}$$

is indefinite. (It has one negative eigenvalue.)

Now (8) calculates the approximate covariances as

$$\hat{R}_0 = 2/3, \qquad \hat{R}_1 = 0, \qquad \hat{R}_2 = \frac{-1}{3}.$$

The matrix is semi-definite

$$\hat{\mathcal{H}}_3 = \begin{pmatrix} 2/3 & 0 & -1/3 \\ 0 & 2/3 & 0 \\ -1/3 & 0 & 2/3 \end{pmatrix} \geq 0.$$

To see that the positive semi-definiteness is preserved with this approximation let

$$\hat{\mathcal{H}}_N = \begin{pmatrix} \hat{R}_0 \hat{R}_1, & \dots, & \hat{R}_{N-1} \\ \hat{R}_1 & & \vdots \\ \vdots & & \vdots \\ \hat{R}_{N-1}, & \dots, & \hat{R}_0 \end{pmatrix}.$$

It can be written in the factored form $(1/N)\hat{Y}\hat{Y}'$ which shows that $\hat{\mathcal{H}}_N \geq 0$,

where

$$\hat{Y} = \begin{bmatrix} 0, \ldots 0, y_0 y_1, \ldots, y_{N-1} \\ \cdot^{\cdot^{\cdot}} \\ y_0, \ldots, y_{N-1}, 0, \ldots, 0 \end{bmatrix} : N \times (2N-1).$$

If the infinite-dimensional matrix T is defined with $\hat{T}$ as its N × N submatrix in its upper left corner and zero everywhere else, and if Y is similary extended so that $T = \frac{1}{N} YY'$, then $T \geq 0$. An alternative proof based on partial realization is given by Kimura [1982].

9 ESTIMATION OF SYSTEM MATRICES: INITIAL PHASE

A Markov model of a weakly stationary time series is constructed by operating on the Hankel matrix made up of the covariances $\Lambda_t = Ey_t y_0'$. In deterministic models, the dimensions of their state space are obtained as the theoretical ranks of the associated Hankel matrices. In stochastic models, rows of Hankel matrices contain noises and the ranks must be determined numerically. Here, system theory has contributed a procedure for approximate model construction by calculating singular values of Hankel matrices, and then properly scaling variables. This second step is known as selecting internally balanced models based on relative sizes of singular values. We also comment on the close relation between the canonical correlation method of Akaike [1976] and the singular value decomposition procedure.

We first describe how to construct full-dimensional models. Then we suggest a method for approximate model construction by examining relative sizes of the singular values of the Hankel matrices. Constructing approximate ARMA or Markov models of low-order this way lets the orders of the approximate models be suggested by data. This property seems to be quite desirable for any model construction method. We later say more on refining the models thus obtained by further optimization steps which maximize the likelihood functions adjusted for the number of parameters used in the models.

9.1 System Matrices

The Hankel matrix in the product form of Chapter 7 shows us a way to construct the system matrices A, B, C in a Markovian representation of time series $\{y_t\}$. Suppose we can construct a Markovian model of a weakly stationary process $\{y_t\}$ as

$$(1)\qquad \begin{cases} \chi_{t+1} = A\chi_t + u_t, \ E\chi_0 = 0, \\ y_t = C\chi_t + v_t, \end{cases}$$

where $\{u_t\}$ and $\{v_t\}$ are mean-zero serially uncorrelated weakly stationary processes with $\mathrm{cov}\begin{pmatrix} u_t \\ v_t \end{pmatrix} = \begin{pmatrix} Q & N \\ N' & R \end{pmatrix}$.

Denote the covariances of $\{y_t\}$ by $\{\Lambda_\ell\}$. They are given by

$$(2)\qquad \Lambda_\ell = Ey_\ell y_0' = CA^{\ell-1}M, \qquad \ell \geq 1, \ \Lambda_0 = C\pi C' + R$$

where

$$M = A\pi C' + N,$$

and

$$\pi = E\chi_0\chi_0'.$$

The weak stationarity and the dynamic equation imply that π satisfies a matrix equation

$$(3)\qquad \pi = A\pi A' + Q.$$

A truncation of the Hankel matrix made up of the covariance matrices Λ's can be written as

$$\mathbb{H}_N = \begin{pmatrix} \Lambda_1 & \Lambda_2 & \cdots & \Lambda_N \\ \Lambda_2 & & \cdots & \Lambda_{N+1} \\ \vdots & & & \\ \Lambda_N & & \cdots & \Lambda_{2N-1} \end{pmatrix} : \ Np\times Np$$

$$= \begin{pmatrix} C \\ CA \\ \vdots \\ CA^{N-1} \end{pmatrix} [M \ \ AM \ \ \ldots \ \ A^{N-1}M].$$

Shift up the submatrices in $\mathbb{H}_N$ by one submatrix row and fill in the bottom submatrix row accordingly to define:

$$\mathcal{H}_A = \begin{pmatrix} \Lambda_2 & \cdots & \Lambda_{N+1} \\ \Lambda_3 & \cdots & \Lambda_{N+2} \\ \vdots & & \\ \Lambda_{N+1} & \cdots & \Lambda_{2N} \end{pmatrix} = \begin{pmatrix} C \\ CA \\ \vdots \\ CA^{N-1} \end{pmatrix} A[M \quad \ldots \quad A^{N-1}M].$$

Take the first submatrix column of $\mathcal{H}_N$ to define

$$\mathcal{H}_M = \begin{pmatrix} \Lambda_1 \\ \vdots \\ \Lambda_N \end{pmatrix} = \begin{pmatrix} C \\ CA \\ \vdots \\ CA^{N-1} \end{pmatrix} M$$

and the first block submatrix row is named

$$\mathcal{H}_C = [\Lambda_1 \quad \ldots \quad \Lambda_N] = C[M, AM, \ldots, A^{N-1}M].$$

The singular value decomposition theorem (see Appendix A.12) tells us that matrices U and V exist such that

$$U'U = I,$$

$$V'V = I,$$

and

$$\mathcal{H}_N = U \Sigma V'$$

where the matrix Σ arranges the singular values of $\mathcal{H}_N$ in decreasing order in magnitude on the main diagonal. Then noting that $\Sigma^{-1/2}U'\mathcal{H}_N V\Sigma^{-1/2} = I_N$, we construct

(4) $$A = \Sigma^{-1/2}U'\mathcal{H}_A V\Sigma^{-1/2}.$$

From the expression for $\mathcal{H}_M$, we construct

(5) $$M = \Sigma^{-1/2}U'\mathcal{H}_M.$$

Similarly, from $\mathcal{H}_C$

(6) $$C = H_C V \Sigma^{-1/2}.$$

These construction steps can be related to the notion of the system matrix of Rosenbrock [1970]. Bosgra and van der Weiden [1980] and van Zee [1981] proposed the procedures followed in this section. See Bosgra and van der Weiden for the proof of these relations.

Arranging the covariance matrices $\{\Lambda_\ell\}$ into the Hankel matrix we can estimate the system matrix A and C of the state space model (1) by (4) and (6). To estimate Π, i.e., the covariance matrix of the initial state vector X_0, we can use (5) if the noise covariance N is known. For example, the relation below (2) can be used to solve for vec Π from vec $M = (C \otimes A)\text{vec}\,\Pi +$ vec N. This matrix Π must, of course, be consistent with (2) and (3), i.e., if Q and R of the noise covariance matrices are known, then Π must satisfy $\Pi = A\Pi A' + Q$ and $\Lambda_0 = C\Pi C' + R$. Once we know A, C and M, and the covariance sequences $\{\Lambda_\ell\}$, then we can estimate Π and the noise covariance matrices Q, N and R by

$$R = \Lambda_0 - C\Pi C'$$

$$N = M - A\Pi C'$$

and

$$Q = \Pi - A\Pi A'$$

where Π is symmetric positive semi-definite and must be such that $\begin{pmatrix} Q & N \\ N' & R \end{pmatrix} \geq 0$. Among all such Π's, we need the minimum Π_* in the ususal partial ordering of symmetric positive definite matices, $\Pi \geq \Pi_*$ because Π_* is associated with the Kalman filter estimates as we later show in Chapter 10. See Faurre [1976] for example. Note also that the model (1) can be replaced by the innovation model we construst in Chapter 10. Nothing of substance changes. We resume our discussion of phase two in Chapter 10 after we first introduce a few more useful concepts related to Hankel matrices.

9.2 Approximate Model

Start with a full dimensional innovation model, $\chi_{t+1} = A\chi_t + Be_t$, $y_t = C\chi_t + e_t$, which is derived in Chapter 10. The idea that relative magnitudes of singular values of the Hankel matrices give us a way to construct approximate models can be quickly grasped by partitioning a state space vector into two subvectors

$$\chi_t = \begin{pmatrix} \chi_t^1 \\ \chi_t^2 \end{pmatrix}$$

where χ_t^1 is assumed to be a lower dimensional approximation to a more complete and higher dimensional vector χ_t. Partition the model conformably and write

$$A = \begin{pmatrix} A_1 & A_{12} \\ A_{21} & A_2 \end{pmatrix}, \quad B = \begin{pmatrix} B_1 \\ B_2 \end{pmatrix}, \quad \text{and } C = (C_1, C_2)$$

in the state space model. Then, $\chi_{t+1}^1 = A_1 x_t^1 + B_1 e_t$, $y_t = C_1 x_t^1 + e_t$ is a lower-dimensional approximation to the model.

The observability matrix $\mathcal{O}$ can be written as

$$\mathcal{O} = [\mathcal{O}_1 + \mathcal{O}_{12}, \mathcal{O}_2]$$

where

$$\mathcal{O}_1 = \begin{bmatrix} C_1 \\ C_1A_1 \\ C_1A_1^2 \\ \vdots \end{bmatrix}.$$

The matrices $\mathcal{O}_{12}$ and $\mathcal{O}_2$ contain everything not explicitly carried by $\mathcal{O}_1$. Similarly, the controllability matrix is written in a partitioned form

$$\mathcal{C} = \begin{pmatrix} \mathcal{C}_1 \\ \mathcal{C}_{21} \end{pmatrix} + \mathcal{C}_2$$

where

$$\mathfrak{C}_1 = [B_1, B_1A_1, B_1A_1^2, \ldots].$$

The matrices $\mathfrak{C}_{21}$ and $\mathfrak{C}_2$ contain terms omitted in $\mathfrak{C}_1$.

Because the Hankel matrix is the product of the observability and controllability matrix, the true $\mathfrak{H}$ is expressible as

$$\mathfrak{H} = \mathfrak{H}^1 + \Delta\mathfrak{H}$$

where $\mathfrak{H}^1 = \mathfrak{O}_1\mathfrak{C}_1$ is the Hankel matrix corresponding to the approximate model, and $\Delta\mathfrak{H}$ contain all other cross-product expressions. The singular value decomposition theorem states that

$$\|\Delta\mathfrak{H}\| = \|\mathfrak{H} - \mathfrak{H}^1\|,$$

where we may use the Frobenius norm of a matrix, $\|X\|^2 = \text{tr } X'X$, i.e., $\|\Delta\mathfrak{H}\| = (\Sigma_{s=r+1}^{n}\sigma_s^2)^{1/2}$, or the spectral norm, $\|\Delta\mathfrak{H}\| = \sigma_{r+1}$. Arrange the singular values of $\mathfrak{H}$ in decreasing order of magnitude. If we decide to have approximation accuracy of σ_{r+1} using the spectral norm, then $\mathfrak{H}^1$ retains the r largest singular values $\sigma_1 \geqslant \ldots \geqslant \sigma_r$, and χ_t^1 becomes r-dimensional. From our discussion on the approximate model construction, to produce an r-dimensional approximate model, partition Σ as diag (Σ_1, Σ_2) where $\Sigma_1 = \text{diag }(\sigma_1, \ldots, \sigma_r)$, U and V conformably; $U = (U_1, U_2)$, $V' = \begin{pmatrix} V_1' \\ V_2' \end{pmatrix}$.

Then

(4') $$A_1 = \Sigma_1^{-1/2}U_1'\mathfrak{H}_A V_1\Sigma_1^{-1/2}$$

(5') $$M_1 = \Sigma_1^{-1/2}U_1'\mathfrak{H}_M$$

and

(6') $$C_1 = \mathfrak{H}_C V_1\Sigma_1^{-1/2}$$

are the system matrices associated with this r-dimensional approximate state space model for $\{y_t\}$.

We can motivate the proposed approximate construction in another way. Because $\mathbb{H}$ represents the input-output characteristics of a dynamic model, an approximation of $\mathbb{H}$ produces an approximate dynamic model. Suppose that r is the dimension of the approximate dynamic model. We must find $\tilde{\mathbb{H}}$ of rank r that best approximates $\mathbb{H}$. The singular value decomposition shows that the Hankel matrix can be put as

$$\mathbb{H}_N = \sum_{i=1}^{n} \sigma_i u_i v_i'$$

where u_i and v_i are the eigenvectors of $\mathbb{H}\mathbb{H}'$ and $\mathbb{H}'\mathbb{H}$ respectively both with eigenvalues σ_i^2, $i = 1, \ldots, n$. If we construct $\mathbb{H}_r = \Sigma_{i=1}^{r} \sigma_i u_i v_i'$, then $\mathbb{H}_r$ minimize $\|\mathbb{H}_N - \mathbb{K}\|$ where $\|\cdot\|$ is the spectral norm among all matrices $\mathbb{K}$ with rank r or less. The minimum equals σ_{r+1}. By construction such an approximation is unique (Kung and Lin [1981]).

This approximating matrix $\mathbb{H}_r$ can be written as

$$\mathbb{H}_r = \mathbb{O}_1 \mathbb{C}_1$$

where

$$\mathbb{O}_1 = U_1 \Sigma^{1/2}$$

where

$$U_1 = [u_1, \ldots, u_r], \qquad \Sigma^{1/2} = \operatorname{diag}(\sigma_1^{1/2}, \ldots, \sigma_r^{1/2}),$$

and

$$\mathbb{C}_1 = \Sigma^{1/2} V_1'$$

where

$$V_1 = [v_1, \ldots, v_r].$$

9.3 Rank Determination of Hankel Matrices: Singular Value Decomposition Theorem

Given a finite number of data vectors from a weakly stationary time series, we now know that the rank of the Hankel matrix is the same as the dimension of a state vector of a Markovian representation of the time series. Because the entries in the Hankel matrix are numerically calculated from the observed data, they have numerical errors associated with them. Numerical determination of the rank of a matrix is ordinarily quite sensitive to errors. We apply the singular value decomposition to the matrix to determine its rank reliably. The singular value decomposition involves only numerically stable procedures.

The singular value decomposition theorem tells us that any m by ℓ matrix can be written as

$$A = U\Sigma V'$$

where

$$U'U = I_m$$

$$V'V = I_\ell$$

and where rank Σ = rank $A = r \leq m, \ell$, in which the submatrix $\Sigma_r = \text{diag}(\sigma_1, \sigma_2, \ldots, \sigma_r)$ is the only non-zero entries in the $(m\times\ell)$ matrix Σ. See Strang [1973].

The following sections describe the way we use this decomposition to construct state space models of time series. We have indicated in the previous section that this decomposition is also used to approximate the state model thus constructed by lower dimensional models, i.e., by state space models with state vectors of lower dimensions. Because $A'AV = V\Sigma^2$, we can interpret V to be the $\ell\times\ell$ matrix made up of ℓ independent eigenvectors of the $\ell\times\ell$ matrix A'A, and $\Sigma^2 = \text{diag}(\sigma_1^2, \ldots, \sigma_r^2, 0 \ldots 0)$ where σ_i^2, $i = 1, \ldots, r$ are the positive eigenvalues of A'A, $i = 1 \ldots r$. Similarly the relation $AA'U = U\Sigma^2$ shows us that the m column vectors in the m m matrix U are the eigenvectors of A'A with $\Sigma^2 = \text{diag}(\sigma_1^2, \ldots, \sigma_r^2, 0 \ldots 0)$. We call σ_i the singular value of A. Appendix A.12 summarizes the other uses of this theorem. We use this decomposition in Section 9 to relate the Hankel matrix method to the canonical correlation method and to the principal component analysis of the covariance matrix.

9.4 Internally Balanced Model

This section examines scalings of variables that go into in the state space representation of time series, and constructs state space models which are numerically well-behaved. We mostly follow Moore [1978], and construct what he calls "internally balanced" models.

Since numerical well-behavedness is a definitely desirable property, constructing (or converting an existing state space model into) an internally balanced model is an important step in the sequence of steps we take to represent time indexed data. The whole process may consist of the following: (i) obtain a state space model of time series either by converting an ARMA model, somehow obtained, into a state space form as in Chapter 5 or by procedures of Sections 1 and 2, (ii) choose an internally balanced model representation by looking for a break in the ordering of singular values of the observability (and controllability) grammians, and (iii) partition the original state vector into two subvectors as suggested in step (ii) to obtain a lower-dimensional approximate model. The Markov model for this subvector is the approximate model which may be converted back into ARMA representation if desired. As an added advantage of this procedure, it generates all approximate models of lower dimensions than the one actually chosen.

<u>Example</u> The next example illustrates the importance of and our concern over scaling. Improper scaling of variables causes some pathological behavior in this example. The impulse response of the system described by

$$\begin{pmatrix} v_{t+1} \\ w_{t+1} \end{pmatrix} = \begin{pmatrix} -1/2 & 0 \\ 0 & -1/3 \end{pmatrix} \begin{pmatrix} v_t \\ w_t \end{pmatrix} + \begin{pmatrix} 10^{-6} \\ 10^{6} \end{pmatrix} u_t,$$

$$y_t = (10^6 \;\; 10^{-6}) \begin{pmatrix} v_t \\ w_t \end{pmatrix},$$

which is denoted by h_i, is equal to $(-1/2)^{i-1} + (-1/3)^{i-1}$, $i = 1, 2, \ldots$ and appears well behaved, showing no obvious anomalies. We note, however, that whatever change in u_t appears enormously magnified on the w variable but it goes nearly unobserved. The opposite ia true with the v variable. In other words, this system is nearly uncontrollable and unobservable because the vectors multiplying u_t in the dynamics and the state vector in the y_t have extreme elements nearly cancelling each other out. This is reflected by the fact that the ellipsoides associated with the controllability and observability grammians are extremely flat. They are defined by $G_o = \Sigma_0^\infty (A')^k C'C(A)^k$, and $G_c = \Sigma_0^\infty A^k BB'(A')^k$. Here, $G_o = G_c = \begin{pmatrix} (4/3)10^{12}, & 6/5 \\ 6/5, & (9/8)10^{-12} \end{pmatrix}$. This example becomes better behaved by a mere rescaling of the components; for example let $\hat{v}_t = 10^6 v_t$, and $\hat{w}_{t+1} = 10^{-6} w_t$. This change of variables produces the model

$$\begin{pmatrix} \hat{v}_{t+1} \\ \hat{w}_{t+1} \end{pmatrix} = \begin{pmatrix} -1/2 & 0 \\ 0 & -1/3 \end{pmatrix} \begin{pmatrix} \hat{v}_t \\ \hat{w}_t \end{pmatrix} + \begin{pmatrix} 1 \\ 1 \end{pmatrix} u_t,$$

$$y_t = (1 \;\; 1) \begin{pmatrix} \hat{v}_t \\ \hat{w}_t \end{pmatrix}.$$

Now the ellipsoides associated with the observability and controllability grammians are $G_o = G_c = \begin{pmatrix} 4/3 & 6/5 \\ 6/5 & 9/8 \end{pmatrix}$, no longer extremely flat.

Let A be asymptotically stable. A state space model (A, B, C) which has the same diagonal matrix Σ as its controllability and observability grammian, i.e., $G_o = G_c = \Sigma$, is called internally balanced, i.e., (A, B, C) is internally balanced if the matrix equations

$$G_c = \sum_0^\infty A^k BB'(A')^k = \Sigma, \quad \Sigma' = \Sigma,$$

and

$$G_o = \sum_0^\infty (A')^k C'CA^k = \Sigma$$

hold. We also speak of an internally balanced representation when a coordinate system in which A, B, and C are represented leads to an internally balanced model. Because the controllability and observability grammians satisfy the matrix algebraic relations $AG_cA' = G_c - BB'$ and $A'G_oA = G_o - C'C$ respectively, the following two equations are simultaneously satisfied by the same diagonal matrix Σ when the system is internally balanced:

(7) $$A\Sigma A' - \Sigma = -BB' \quad \text{and} \quad A'\Sigma A - \Sigma = -C'C.$$

Construction

The internally balanced representation can be constructed by following a two-step procedure (Moore [1978]). The idea is related to the principal component analysis in statistics. We return to this connection later in this section. Also, see Appendix A.3. Let the controllability grammian for the system (A, B, C) be G_c. The matrix G_c has an orthonormal eigenvector matrix Γ_c and the diagonal eigenvalue matrix Δ_c, i.e., $G_c\Gamma_c = \Gamma_c\Delta_c$ or $G_c = \Gamma_c\Delta_c\Gamma_c'$.

(i) Change the coordinate so that $\hat{A} = P^{-1}AP$, $\hat{B} = P^{-1}B$, and $\hat{C} = CP$ where we choose P to be $\Gamma_c\Delta^{1/2}$. Then $\hat{G}_c = P^{-1}G_c(P^{-1})' = \Delta^{-1/2}\Gamma_c'\Gamma_c\Delta_c\Gamma_c'\Gamma_c\Delta_c^{-1/2} = I$. The observability grammian becomes $\hat{G}_o = P'G_oP$. Let $\hat{\Gamma}_o$ and $\hat{\Delta}_o$ be the eigenvector and eigenvalue matrices of G_o: $\hat{G}_o\hat{\Gamma}_o = \hat{\Gamma}_o\hat{\Delta}_o$.

(ii) Perform another change of variables so that

$$\tilde{A} = Q^{-1}\hat{A}Q, \quad \tilde{B} = Q^{-1}\hat{B}, \quad \text{and } \tilde{C} = \hat{C}Q$$

where

$$Q = \hat{\Gamma}_o\hat{\Delta}_o^{-1/4}.$$

The controllability grammian becomes $\tilde{G}_c = Q^{-1}\hat{G}_c(Q^{-1})' = \hat{\Delta}_o^{1/2}$, and the observability grammian becomes $\tilde{G}_o = Q'\hat{G}_oQ = \hat{\Delta}_o^{-1/4}\hat{\Gamma}_o'\hat{\Gamma}_o\hat{\Delta}_o\hat{\Gamma}_o'\hat{\Gamma}_o\hat{\Delta}_o^{-1/4} = \hat{\Delta}_o^{1/2}$, completing the conversion to the internally balanced representation.

Recalling our discussion on the singular value decomposition of the Hankel matrix, we now show that the matrix Σ there is the same as the Σ we have introduced in internally balanced representation. To see this, recall that the Hankel matrix has the factored form $\mathcal{H} = \mathcal{O}\mathcal{C}$. Hence $\mathcal{H}'\mathcal{H} = \mathcal{C}'\mathcal{O}'\mathcal{O}\mathcal{C}$. The product of the observability grammian with the controllability grammian produces $G_oG_c = \mathcal{O}'\mathcal{O}\mathcal{C}\mathcal{C}'$. Suppose G_o and G_c are both positive definite. We now show that nonzero eigenvalues of $\mathcal{H}'\mathcal{H}$ are the eigenvalues of G_oG_c. Let u be an eigenvector of G_oG_c with λ as its eigenvalue, $G_oG_cu = \lambda u$. Then $\mathcal{C}'G_oG_cu = \mathcal{H}'\mathcal{H}\mathcal{C}'u = \lambda\mathcal{C}'u$, i.e., $\mathcal{C}_n'u$ is an eigenvector of $(\mathcal{H}'\mathcal{H})_n$ with the same eigenvalue where $(\ )_n$ denotes $(n \times n)$ submatrix. Conversely, start from $(\mathcal{H}'\mathcal{H})_nu = \lambda u$. This equation implies that $(\mathcal{O}\mathcal{C})_nu = \lambda(\mathcal{C}'\mathcal{O}')_n^{-1}u$ which equals $\mathcal{O}_n'^{-1}\mathcal{C}_n'^{-1}$ if the system is controllable and observable because rank $\mathcal{O}_n'$ = rank $\mathcal{C}_n'$ = n. Hence $\mathcal{O}_n'(\mathcal{O}\mathcal{C})_nu = \lambda\mathcal{C}_n'^{-1}u$. Let $u = \mathcal{C}_n'v$ to rewrite it as $(\mathcal{O}'\mathcal{O}\mathcal{C}\mathcal{C}')_nv = G_oG_cv = \lambda v$, i.e., v is an eigenvector of G_oG_c with the same eigenvalue λ of $(\mathcal{H}'\mathcal{H})_n$. The above shows that if $\sigma_1^2 \geq \sigma_2^2 \geq \ldots \geq \sigma_n^2$ are the eigenvalues of G_oG_c, then $\sigma_1 \ldots \sigma_n$ are the singular values of $\mathcal{H}$. Because the Hankel matrix $\mathbb{H}$ related to (A, B, C) is expressible as $\mathbb{O}\mathbb{C}$, the matrix $\mathbb{H}\mathbb{H}'$ equals $\mathbb{O}\mathbb{C}\mathbb{C}'\mathbb{O}'$ or $\mathbb{O}G_c\mathbb{O}'$. In an internally balanced representation, $\mathbb{H}\mathbb{H}'$ then equals $\mathbb{O}\Sigma\mathbb{O}'$ where Σ is a diagonal matrix. From $\mathbb{H}\mathbb{H}'\mathbb{O} = \mathbb{O}\Sigma\mathbb{O}'\mathbb{O} = \mathbb{O}\Sigma G_o = \mathbb{O}\Sigma^2$, where we use $\mathbb{O}\mathbb{O}' = G_o$ with the same Σ as in an internally balanced representation. We conclude then that the elements of Σ^2 are the eigenvalues of $\mathbb{H}\mathbb{H}'$ or the squares of the singular values of $\mathbb{H}$.

By examining the controllability and observability matrix

$$\tilde{\mathbb{C}} = [\tilde{B}, \widetilde{AB}, \ldots] = Q^{-1}P^{-1}[B, AB, \ldots] = Q^{-1}P^{-1}\mathbb{C}$$

and

$$\tilde{\mathbb{O}} = \begin{bmatrix} \tilde{C} \\ \widetilde{CA} \\ \vdots \end{bmatrix} = \begin{bmatrix} C \\ CA \\ \vdots \end{bmatrix} PQ = \mathbb{O}PQ$$

we note that the Hankel matrices are related by

$$\tilde{\mathbb{H}} = \tilde{\mathbb{O}}\tilde{\mathbb{C}} = \mathbb{O}PQQ^{-1}P^{-1}\mathbb{C} = \mathbb{H}.$$

This is to be expected because the Hankel matrices provide an external description of dynamics and is invariant with respect to basis choices to represent dynamics.

Properties of Internally Balanced Models*

When the two equations in (6) determining the controllability and observability grammians of an internally balanced model are combined, the grammian Σ satisfies an algebraic matrix equation

$$A'A\Sigma A'A - \Sigma = -(C'C + A'BB'A).$$

Let v be an eigenvector of $A'A$ with its corresponding eigenvalue λ. Then the above equation yields, on multiplication by v' from the left, and by v from the right,

---

* This section follows Pernabo and Silverman [1982].

$$(\lambda^2 - 1)v'\Sigma v = -v'(C'C + A'BB'A)v \leq 0,$$

establishing that the eigenvalue of A'A are less than or equal to one in modulus. A further refinement of the argument can show that $|\lambda| < 1$ if the eigenvalues of Σ are all distinct (Pernabo and Silverman [1982]).

Consider a partition of an internally balanced model into two subsystems

$$\chi_{t+1} = \begin{pmatrix} A_{11} & A_{12} \\ A_{21} & A_{22} \end{pmatrix} \chi_t + \begin{pmatrix} B_1 \\ B_2 \end{pmatrix} u_t$$

and

$$y_t = [C_1, C_2]\chi_t.$$

The associated controllability and observability grammians both become block diagonal, $\Sigma = \text{diag}(\Sigma_1, \Sigma_2)$. Suppose that the total system is asymptotically stable, i.e., $\|A\| < 1$. From the construction of an internally balanced representation, we know that $G_0 = G_c = \Sigma$. Assume that Σ is nonsingular. We now establish that every subsystem of an asymptotically stable internally balanced model is asymptotically stable. First, using the defining relation

$$A_{11}\Sigma_1 A'_{11} + A_{12}\Sigma_2 A'_{12} - \Sigma_1 = -B_1 B'_1$$

and by multiplying it by v' and v from left and right respectively, we deduce

$$(|\lambda|^2 - 1)v'\Sigma_1 v = -(v'A_{12}\Sigma_2 A'_{12}v + v'B_1 B'_1 v) \leq 0,$$

where v is now redefined to satisfy $A_{11}v = \lambda v$, $v'v = 1$. Because $v'\Sigma_1 v > 0$, it easily follows that $|\lambda| \leq 1$. We can exclude the possibility that $|\lambda| = 1$ because then $v'A_{12} = 0$ and $v'B_1 = 0$ must follow because Σ_2 is positive definite. But this implies that

$$(v', 0)\begin{pmatrix} A_{11} & A_{12} \\ A_{21} & A_{22} \end{pmatrix} = \lambda(v', 0) \text{ and } (v', 0)\begin{pmatrix} B_1 \\ B_2 \end{pmatrix} = 0$$

hence the system is not reachable, contrary to our assumption. We conclude then that $|\lambda| < 1$ and the subsystem 1 is asymptotically stable. Since subsystem 1 is any subsystem, subsystem 2 is also asymptotically stable.

Suppose we partition the total system according to the criterion $\sigma_{min}(\Sigma_1) > \sigma_{max}(\Sigma_2)$ and that the observability grammian Σ is diagonal. For the subsystem 1, Σ_1 satisfies

$$A_{11}'\Sigma_1 A_{11} + A_{21}'\Sigma_2 A_{21} - \Sigma_1 = -C_1'C_1.$$

If the subsystem 1 is not observable, there is a normalized eigenvector of A_{11}, v, v'v = 1 satisfying $A_{11}v = \lambda v$, and $C_1 v = 0$. Multiplying the above equation by v' and v from the left and right, respectively, it becomes

(8) $$(1 - |\lambda|^2)v'\Sigma_1 v = v'A_{21}'\Sigma_2 A_{21}v.$$

Note that $v'\Sigma_1 v \geq \sigma_{min}(\Sigma_1)$. We can also bound the right hand side by

$$v'A_{21}'\Sigma_2' A_{21}v \leq \|A_{21}v\|^2 \sigma_{max}(\Sigma_2).$$

Internally balanced models are such that $\|A\| \leq 1$. This implies in particular

$$\left\| \begin{pmatrix} A_{11} \\ A_{21} \end{pmatrix} v \right\| \leq 1 \qquad \text{or} \qquad \|A_{11}v\|^2 + \|A_{21}v\|^2 \leq 1$$

hence $\|A_{21}v\|^2 \leq 1 - |\lambda|^2$. Substituting these into (8), we obtain

$$(1 - |\lambda|^2)\sigma_{min}(\Sigma_1) \leq (1 - |\lambda|^2)\sigma_{max}(\Sigma_2).$$

Previous results show that $|\lambda| < 1$, hence $\sigma_{min}(\Sigma_1) \leq \sigma_{max}(\Sigma_2)$. This contradicts the assumed criterion for partitioning subsystems. Hence we conclude that subsystem 1 is observable. Proceeding analogously we also establish that subsystem 1 is also reachable. Kung and Lin [1981] also discuss a model reduction method using the singular value decomposition.

Principal Component Analysis

The notion of internal balanced model corresponds to that of principal components in statistics. Principal components are defined for a p-dimensional random vector x with mean 0 and covariance matrix X. Because X is symmetric and positive semi-definite, p normalized eigenvectors are used to define a p×p orthonormal matrix Γ with $X\Gamma = \Gamma\Lambda$, where Λ is the diagonal matrix made up of of the eigenvalues. By definition, the largest p-components of $\Gamma' x$ are the p principal components of x. The first principal component is produced by $\gamma_1' x$ where γ_1 is the normalized eigenvector corresponding to the largest eigenvalue λ_1.

We note that the coordinate changes used to construct internally balanced models calculate principal components:* In going from (A, B, C) to $(\hat{A}, \hat{B}, \hat{C})$, the new state vector z_t is related to the old one by $\hat{z}_t = P^{-1} z_t = \Delta^{-1/2} \Gamma_c' z_t$. The components of $\Delta^{1/2} \hat{z}_t$ are exactly the principal components of z_t. Similarly, the step from $(\hat{A}, \hat{B}, \hat{C})$ to $(\tilde{A}, \tilde{B}, \tilde{C})$ involves the change of variables $\tilde{z}_t = Q^{-1} \hat{z}_t = \hat{\Delta}_0^{1/4} \Gamma_0' \hat{z}_t$, i.e., aside from scaling, $\hat{\Gamma}_0' \hat{z}_t$ calculates the principal components of $\hat{z}_t$.

The relation $\operatorname{tr} \Gamma' X \Gamma = \operatorname{tr} X \Gamma \Gamma' = \operatorname{tr} X$ shows that the total variance of the principal components is the same as that of x, i.e., $\Sigma_{i=1}^{p} \lambda_i$, and the first principal component explains λ_1 of $\Sigma_1^p \lambda_i$. Thus, if $\lambda_2 + \ldots + \lambda_p$ is small compared with λ_1, the first principal component explains most of the variation of x.

This static definition can formally be extended to dynamic situations. Suppose y_t is produced from a white noise sequence $\varepsilon_t, \varepsilon_{t-1}, \ldots$ by $y_{t+1} = \mathbb{C} u_t$ where $\mathbb{C} = [B, AB, \ldots]$ and $u_t' = (\varepsilon_t, \varepsilon_{t-1}, \ldots)$. Assume that ε_t is mean zero and $E(\varepsilon_t \varepsilon_s') = I \delta_{t,s}$. Then $\operatorname{cov}(y_{t+1}) = \mathbb{C}\mathbb{C}' = G_c$. Then $\Gamma' y_t$ is the vector of principal components where $G_c \Gamma = \Gamma \Lambda$ and Λ is the diagonal matrix of the eigenvalues of G_c. In an internally balanced model G_c is already diagonal; hence the first component of y_t is its principal component.

9.5 Inference about the Model Order

How do we let data determine the order of an ARMA model? The answer

---

* Appendix A.3 summarizes relevant properties of the principal components.

largely depends on our a priori belief in the appropriateness or correctness of a class of ARMA models. If we firmly believe in ARMA models as correct representations of dynamic mechanisms generating data, then consistency of estimates of the order and the system parameters should be our primary concern. If, on the other hand, ARMA models are used merely as convenient approximate representations of much more complex dynamic phenomena, then we have no compelling reason to emphasize consistency of estimates. We would rather strive to balance bias of estimates and loss of efficiency from employing too many system parameters, and try to achieve asymptotically efficient approximations to the true spectrum of the underlying process. Deistler et al. [1982] express a similar opinion.

From the former standpoint, a criterion

$$\mathrm{BIC}(p, q) = \ln \sigma(p, q) + (p+q)\ln N/N$$

has been proposed by Rissanen [1976, 1983], Akaike [1973, 1976] and Schwarz [1978], where p+q is the total number of parameters and $\sigma(p, q)$ is the standard deviation of the innovation process. Adopting the latter point of view, Akaike proposed

$$\mathrm{AIC}(p, q) = \ln \sigma(p, q) + 2(p+q)/N$$

as a criterion to choose the order the model ARMA(p, q).

Suppose that data Y_i, $i = 1, \ldots, N$ are independent random variables with density p(y). This AIC criterion may be regarded as an approximation that minimizes the distance between the true probability density function p(y) and its best approximation $f(y, \hat{\theta})$ chosen from a class of functions $f(y, \theta)$, where θ is a finite dimensional vector and $\hat{\theta}$ is its estimate. Takeuchi [1983] explains the approximations used to derive the AIC criterion.

9.6 Choices of Basis Vectors

A transfer function has infinitely many equivalent representations in state space from. Let $\underline{S}_n$ denote the set of all possible state space models of dimension n and $\underline{S}_n/\sim$ denote its equivalence class. A function defined on $\underline{S}_n$ is called an invariant for "~" if its value is the same for two equivalent representations, i.e., if x, $y \in \underline{S}_n/\sim$, then f(x) = f(y). The Markov parameters are shown to be invariants in Chapter 7.

Let us now return to the Hankel matrix (4) associated with conditional predictions of future y's of Section 7.1. We can assume that the p components of the vector $y_{t+1|t}$ are all linearly independent. The first p row vectors of $\mathcal{H}$ are then automatically included in any basis for the row space of $\mathcal{H}$. Because the rows of $\mathcal{H}$ are made up of blocks of p row vectors and because of the regular patterns of submatrices of $\mathcal{H}$, if row i is in the linear span of the preceding rows, then so is row (i + p).

A basis for the row space of $\mathcal{H}$ is denoted by a set of n indeces that designate the row vectors in the basis, $\underline{i}$, $i_1 < i_2 < \dots < i_n$, where n = rank $\mathcal{H}$. Number the rows of $\mathcal{H}$ by h_{jk}, where h_{jk} denotes the j-th component of $y_{t+k|t}$. A basis $\underline{i}$ can be specified by p integers (called structure indices): The set $\{n_1, n_2, \dots, n_p\}$, where $\Sigma_i n_i = n$, and n_k is the smallest integer such that the row $k + n_k p$ is not in the basis, k = 1 ... p. This construction means that the rows $h_{11} \dots h_{1n_1}, h_{21} \dots h_{2n_2}, \dots, h_{p1} \dots h_{pn_p}$ are in the basis $\underline{i}$. To select basis vectors, start from h_{11}. Next check the first elements of h_{12}, h_{13}, etc., and stop at h_{1n_1}, the first element of the n_1-th block (first component of $y_{t+n_1-1|t-1}$). Then $h_{1,\,n_1+1}$ is linearly dependent on the rows previously chosen. This ends the first string. Next we start with h_{21}, and so forth.

A simple alternative selection rule of basis vectors that turns out to be sensitive to numerial noises such as round-offs, or sampling error is this: Choose the first n linearly independent vectors. This selection rule leads to a state space model in cannonical form as we later explain.

An example may clarify the procedure. Suppose $n = 3$ and $p = 2$. The next table shows $\underline{i} = (1,2,4)$ where x denotes independent rows and 0 dependent rows. Here $n_1 = 1$ and $n_2 = 2$ because h_{11}, h_{21}, and h_{22} are in the basis.

	Row #		1st string	2nd string
$y_{t\|t-1}$	1	h_{11}	x	
	2	h_{21}		x
$y_{t+1\|t-1}$	3	h_{12}	0 (n_1=1)	
	4	h_{22}		x
$y_{t+2\|t-1}$	5	h_{13}	0	
	6	h_{23}		0 (n_2=2).

In the cannonical selection rule, the row vectors are examined in sequence, h_{11}, h_{21}, h_{12}, h_{22} etc. In this model the same set of rows h_{11}, h_{21}, h_{22} is selected.

The next table illustrates a possible basis for $i = (1, 2, 3, 4, 6)$, $n = 5$ and $p = 3$.

	Row #		1st string	2nd string	3rd string
$y_{t\|t-1}$	1	h_{11}	x		
	2	h_{21}		x	
	3	h_{31}			x
$y_{t+1\|t-1}$	4	h_{12}	x		
	5	h_{22}		0	
	6	h_{32}			x
$y_{t+2\|t-1}$	7	h_{13}	0		
	8	h_{23}		0	
	9	h_{33}			0
			(n_1=2)	(n_2=1)	(n_3=2)

The basis vectors are selected in a member of "passes" in both tables. In the first table, the first string of linearly independent row vectors, which are selected in the first pass over the rows of $\mathcal{H}$, consists of h_{11} by itself because h_{12} is linearly dependent on h_{11}. We earlier remarked that if h_{21} is linearly dependent so are all h_{j1}, $j = 3, 4, \ldots$. The second pass yields h_{21} and h_{22} because h_{23} becomes linearly dependent of the previously chosen basis vectors. In the second table, the third string contains h_{31}. The vector h_{33} is linearly dependent on the previously chosen vectors, hence $n_3 = 2$.

In the case of the first table, if $(n_1 = 1, n_2 = 2)$ resulted from choosing the first three linearly independent vectors, then

$$h_{12} = \alpha_{111}h_{11} + \alpha_{121}h_{21}$$

and

$$h_{23} = \alpha_{211}h_{11} + \alpha_{221}h_{21} + \alpha_{222}h_{22}.$$

But if $(n_1 = 1, n_2 = 2)$ resulted not by choosing the first three linearly independent vectors, then

$$h_{12} = \alpha_{111}h_{11} + \alpha_{121}h_{21} + \alpha_{122}h_{22}$$

would result.

The former expression for h_{12} is then the canonical representation. By construction of the structure indices,

(9) $$h_{i,\, n_i+1} = \sum_{j=1}^{p} \sum_{k=1}^{n_j} \alpha_{ijk}\, h_{jk}, \quad i = 1 \ldots p.$$

Equivalently, $y^i_{t+n_1|t-1} = \sum_{j=1}^{p} \sum_{k=1}^{n_j} {}_{ijk}\, y^j_{t+k-1|t-1}$.

These np numbers $\{\alpha_{ijh}\}$ and another np numbers, the first p elements of the row vector h_{ij} $i = 1 \ldots p$, $j = 1 \ldots n_i$, $k = 1 \ldots p$, together completely specify the Markov parameters. The latter np parameters appear in constructing state dynamic equations as we now show.

9.7 State Space Model

Now, define a state vector by

$$\chi'_{t+1} = [y^1_{t+1|t}, y^1_{t+2|t}, \ldots, y^1_{t+n_1|t}, \ldots, y^p_{t+1|t}, \ldots, y^p_{t+n_p|t}],$$

where $y^j_{t+k|t}$ is the j-th component of the vector $y_{t+k|t}$ and corresponds to $h_{j,k+1}$. Take the conditional expectation of the representation of $\{y_t\}$ given by $y_{t+k} = \Sigma_0^\infty H_{i|t+k-i}$ with respect to the information set at t to obtain

$$y_{t+k|t} = \sum_{i=k}^{\infty} H_i \varepsilon_{t+k-i} = H_k \varepsilon_t + \sum_{i=k+1}^{\infty} H_i \varepsilon_{t+k-i}$$

or recognizing that the second term equals $y_{t+k|t-1}$, we can write

$$y_{t+k|t} = H_k \varepsilon_t + y_{t+k|t-1}.$$

We can thus express the state vector χ_{t+1} as

$$\chi_{t+1} = \begin{pmatrix} y^1_{t+1|t-1} \\ \vdots \\ y^p_{t+n_p|t-1} \end{pmatrix} + K\varepsilon_t \quad \text{where } K = \begin{pmatrix} h_{11}(1) & \ldots & h_{11}(p) \\ \vdots & & \vdots \\ h_{1n_1}(1) & \ldots & h_{1n_1}(p) \\ \vdots & & \vdots \\ h_{pn_p}(1) & \ldots & h_{pn_p}(p) \end{pmatrix}.$$

In view of (9) the first term can be written as $F\chi_t$ if the matrix F is introduced with the block diagonal structure

$$F = \begin{pmatrix} F_{11} & F_{12} & \cdots & F_{1p} \\ \vdots & & & \\ \vdots & & & \\ F_{p1} & \cdots & \cdots & F_{pp} \end{pmatrix}$$

where the diagonal submatrices are

$$F_{ii} = \begin{pmatrix} 0 & & & \\ \vdots & & I_{n_i-1} & \\ \vdots & & & \\ \alpha_{i11} & \alpha_{i12} & \cdots & \alpha_{iln_i} \end{pmatrix} \qquad i = 1, \ldots, p$$

and off-diagonal submatrices are similar except for the absence of the identity matrix block,

$$F_{12} = \begin{pmatrix} 0 & \dots & 0 \\ & & \cdot \\ \alpha_{121} & \dots & \alpha_{12n_2} \end{pmatrix}, \text{ for example.}$$

The α's in the last rows of F_{ij} are the system parameters. Because $y_t = y_{t|t-1} + \varepsilon_t$, the prediction $y_{t|t-1}$ can be reconstructed from χ_t by picking up the first components of each of p blocks, i.e., $y_{t|t-1} = N\chi_t$ where $N = [N_1, \dots, N_p]$ are simply given by

$$N_1 = \begin{pmatrix} 1 & 0 & \dots & 0 \\ \vdots & \vdots & & \vdots \\ 0 & 0 & \dots & 0 \end{pmatrix}, \dots, N_p = \begin{pmatrix} 0 & 0 & \dots & 0 \\ \vdots & \vdots & & \vdots \\ 1 & 0 & \dots & 0 \end{pmatrix}.$$

Example (Denham [1974]) Let p = 2 and n = 1. Because the dimension is one, only one row vector is linearly independent, so $\underline{i}$ contains a single element. We consider three alternative choices of the index set: $i = \{1\}$, $i = \{2\}$, and $i = \{3\}$.

$$i = \{1\}: \quad y^1_{t+1|t} = \alpha_1 y^1_{t|t-1} + [h_{11}(1), h_{11}(2)]\varepsilon_t,$$

$$y_t = \begin{pmatrix} 1 \\ \beta_{12} \end{pmatrix} y^1_{t|t-1} + \varepsilon_t;$$

$$i = \{2\}: \quad y^2_{t+1|t} = \alpha_2 y^2_{t|t-1} + [h_{21}(1), h_{21}(2)]\varepsilon_t,$$

$$y_t = \begin{pmatrix} \beta_{21} \\ 1 \end{pmatrix} y^2_{t|t-1} + \varepsilon_t;$$

$$i = \{3\}: \quad y^1_{t+2|t} = \alpha_3 y^1_{t+1|t-1} + [h_{12}(1), h_{12}(2)]\varepsilon_t,$$

$$y_t = \begin{pmatrix} \beta_{31} \\ \beta_{32} \end{pmatrix} y^1_{t+1|t-1} + \varepsilon_t.$$

The canonical form corresponding to $i = \{2\}$ is

$$y^2_{t+1|t} = \alpha_2 y^2_{t|t-1} + [h_{21}(1), h_{21}(2)]\varepsilon_t,$$

$$y_t = \begin{pmatrix} 0 \\ 1 \end{pmatrix} y^2_{t|t-1} + \varepsilon_t,$$

because $y^2_{t|t-1}$ being the first linearly independent vector implies that $y^1_{t|t-1}$ is not, i.e., $y^1_{t|t-1} = 0$.

Except for the nongeneric case $\beta_{12} = 0$ or $\beta_{21} = 0$, either form can parameterize all possible processes except for the canonical form.

Another example, from Wertz [1981], also illustrates the special nature of the canonical representation. Let $p = 2$, first with $n = 3$ and $n_1 = 2$ and $n_2 = 1$ so that $\underline{i}_1 = (1, 2, 3)$ and then with $n_1 = 1$, $n_2 = 2$ so that $\underline{i}_2 = (1, 2, 4)$. In the first case $\chi_{t+1} = (y^1_{t+1|t}, y^1_{t+2|t+1}, y^2_{t+1|t})'$. The state dynamic equation becomes

$$\chi_{t+1} = \begin{pmatrix} 0 & 1 & 0 \\ \alpha_{111} & \alpha_{112} & \alpha_{121} \\ \alpha_{211} & \alpha_{212} & \alpha_{221} \end{pmatrix} \begin{pmatrix} y^1_{t|t-1} \\ y^1_{t+1|t} \\ y^2_{t|t-1} \end{pmatrix} + \begin{pmatrix} h_{11}(1) h_{11}(2) \\ h_{12}(1) h_{12}(2) \\ h_{21}(1) h_{21}(2) \end{pmatrix} \varepsilon_t$$

and

$$y_t = \begin{pmatrix} 1 & 0 & 0 \\ 0 & 0 & 1 \end{pmatrix} \chi_t + \varepsilon_t .$$

In the second case, the state vector is $w_{t+1} = (y^1_{t+1|t}, y^2_{t+1|t}, y^2_{t+2|t+1})'$. The Markovian model becomes

$$w_{t+1} = \begin{pmatrix} \alpha_{111} & \alpha_{121} & \alpha_{122} \\ 0 & 0 & 1 \\ \alpha_{211} & \alpha_{221} & \alpha_{222} \end{pmatrix} \begin{pmatrix} y^1_{t|t-1} \\ y^2_{t|t-1} \\ y^2_{t+1|t} \end{pmatrix} + \begin{pmatrix} h_{11}(1) h_{21}(2) \\ h_{21}(1) h_{21}(2) \\ h_{22}(1) h_{22}(2) \end{pmatrix} \varepsilon_t$$

$$y_t = \begin{pmatrix} 1 & 0 & 0 \\ 0 & 1 & 0 \end{pmatrix} w_t + \varepsilon_t .$$

If $\underline{i} = (1, 2, 4)$ is the set of the first three linearly independent vectors, then α_{122} is zero yielding the canonical form.

The representation for $\underline{i} = (1, 2, 3)$ and $\underline{i} = (1, 2, 4)$ overlaps if the latter does not correspond to the canonical form because the transformation

$$\chi_t = \begin{pmatrix} 1 & 0 & 0 \\ \alpha_{111} & \alpha_{121} & \alpha_{122} \\ 0 & 1 & 0 \end{pmatrix} w_t$$

changes the representation from one local coordinate $\underline{i}_1$ to $\underline{i}_2$ in all generic cases, i.e., when $\alpha_{122} \neq 0$. Note that this is not possible for the canonical form because the transformation is singular if α_{122} is zero.

9.8 ARMA (Input-Output) Model

A choice of the basis for the row space of $\mathcal{H}$ equivalently yields ARMA models. Suppose that the structure indices are $n_1, \ldots, n_p$. The dependent rows in $\mathcal{H}$ are given by (1) of the previous section. It is reproduced here for easy reference.

$$y^i_{t+n_i|t-1} = \sum_{j=1}^{p} \sum_{k=1}^{n_j} \alpha_{ijh} y^j_{t+k-1|t-1}, \qquad i = 1, \ldots, p.$$

Write this for all p series as

$$(+) \qquad A(z) y_{t|t-1} = 0$$

where the elements of A(z) are:

$$a_{ii}(z) = z^{n_i} - \alpha_{iin_i} z^{n_j - 1} - \ldots - \alpha_{ii1}$$

$$a_{ij}(z) = -\alpha_{ijn_j} z^{n_j - 1} - \ldots - \alpha_{ij1}, \qquad i \neq j.$$

Let r be the maximum of the structure indices, max $(n_1, \ldots, n_p)$. Then (2) can equivalently be written as

$$A_0 y_{t+r|t-1} + A_1 y_{t+r-1|t-1} + \ldots + A_r y_{t|t-1} = 0.$$

Noting that $y_t = y_{t|t-1} + \varepsilon_t$, and

$$y_{t+k|t-1} = \sum_{k-1}^{\infty} H_i \varepsilon_{t+k-i} = y_{t+k} - \ldots - H_0 \varepsilon_{t+k} - \ldots - H_k \varepsilon_t,$$

the ARMA model corresponding to (+) is obtained:

$$A_0 y_{t+r} + A_1 y_{t+r-1} + \ldots + A_r y_t = B_0 \varepsilon_{t+r} + \ldots + B_r \varepsilon_t$$

where

$$B_0 = A_0$$

$$B_1 = A_1 + A_0 H_1$$

$$\vdots$$

$$B_r = A_r + \ldots + A_1 H_{r-1} + A_0 H_r.$$

By construction, $\deg a_{ij}(z) < \deg a_{jj}(z) = n_j$, and

$$\deg \ell_{ij}(z) = \begin{cases} r \quad , \text{ if } n_j = r, \\ r-1, \text{ if } n_j < r. \end{cases}$$

The parameters of A(z) are exactly the same as the np parameters α_{ijk} in the state transition matrix F. The matrix B(z) as calculated above contains row of not in the state space model.

To see that B(z) can be reduced, start from the state space model constructed in the previous section. Partition χ_t comformably into p subvector components

$$\chi_t = \begin{pmatrix} \chi_{1t} \\ \chi_{2t} \\ \vdots \\ \chi_{pt} \end{pmatrix}.$$

From the special structure of the matrix N in $y_t = Nz_t + \varepsilon_t$, y_t^i equals the first vector of z_{jt} plus ε_t, i.e.,

$$\chi_{jt}^1 = y_t^j - \varepsilon_t^j.$$

The dynamics $\chi_{t+1} = F\chi_t + K\varepsilon_t$ show that

$$\chi_{jt+1}^1 = \chi_{jt}^2 + k_{j1}\varepsilon_t$$

or

$$\chi_{jt}^2 = y_{t+1}^j - \varepsilon_{t+1}^j - k_{j1}\varepsilon_t = zy_t^i - z\varepsilon_t^j - k_{j1}\varepsilon_t$$

where $k_{j1} = (h_{j1}(1), \ldots, h_{j1}(p))$.

Proceeding in the same way

$$\chi_{jt+1}^2 = \chi_{jt}^3 + k_{j2}\varepsilon_t$$

or

$$\begin{aligned} \chi_{jt}^3 &= \chi_{jt+1}^2 - h_{j2}\varepsilon_t \\ &= y_{t+2}^j - \varepsilon_{t+2}^i - k_{j1}\varepsilon_{t+1} - k_{j2}\varepsilon_t \\ &= z^2 y_t^i - z^2\varepsilon_t^j - k_{j1}z\varepsilon_t - k_{j2}\varepsilon_t, \end{aligned}$$

$$\vdots$$

$$\chi_{jt}^{nj} = z^{n_j-1} y_t^i - z^{n_j-1} \varepsilon_t^j - k_{jn_j-1}\varepsilon_t - \ldots - k_{j1} z^{n_j-2} \varepsilon_t.$$

Collecting them together, we can express χ_t as

(10) $\chi_t = V(z)y_t - w(z)\varepsilon_t$

where

$$V(z) = \begin{pmatrix} V_1(z) & 0 & \ldots & 0 \\ 0 & V_2(z) & & \vdots \\ 0 & 0 & & 0 \\ \vdots & \vdots & & \vdots \\ 0 & 0 & & V_p(z) \end{pmatrix}$$

$$V_i(z) = \begin{pmatrix} 1 \\ z \\ \vdots \\ z^{ni} \end{pmatrix} \qquad i = 1, \ldots, p,$$

and

$$W(z) = \begin{pmatrix} W_1(z) & 0 & \ldots & 0 \\ 0 & W_2(z) & \ldots & 0 \\ 0 & \ldots & \ldots & W_p(z) \end{pmatrix}.$$

9.9 Canonical Correlation

Earlier we have introduced Hankel matrices as a way of relating the predicted future realization of a time series to the realization of the current and past exogenous noises. Another Hankel matrix results from calculating the correlation of future realizations of a time series with a finite set of past data

$$E\begin{pmatrix} y_{t+1} \\ y_{t+2} \\ \vdots \\ y_{t+N} \end{pmatrix}[y_t' \;\; y_{t-1}' \cdots y_{t-N+1}'] = \begin{pmatrix} \Lambda_1 & \Lambda_2 & \cdots & \Lambda_N \\ \Lambda_2 & \Lambda_3 & \cdots & \Lambda_{N+1} \\ \vdots & & & \\ \Lambda_N & \Lambda_{N+1} & \cdots & \Lambda_{2N-1} \end{pmatrix}$$

where

$$\Lambda_\ell = E(y_{\ell+1}y_1').$$

This idea of projecting future observations on the subspace in the Hilbert space spanned by the current and past noises or observations is a basis for many algorithms for prediction and model construction (the so-called stochastic realization problem). For example, see Faurre [1976] or Akaike [1976] to which we shortly return.

In practice, a sequence of sample covariances $\hat{R}_t = (1/N)\Sigma_{i=1}^{N-t} y_{t+s}y_s'$ is constructed from a data set $\{y_1 \ldots y_{N-t}\}$.* Then, for numerical reasons, the Hankel matrix H_p is often scaled by $R_p^{-1/2}H_pR_p^{1/2}$, where $R_p = (R_{|i-j|})$, $1 \leq i$, $j \leq p$ is a block Toeplitz matrix with the same submatrices arranged on block diagonal lines, and the singular value decomposition applied to the rescaled matrix.

If ν is an upper bound on the order of the ARMA model, then the $(\nu\times\nu)$ upper left hand corner block submatrix of the Hankel matrix can be generated as the correlation matrix of the future and past data vectors: Define

$$Y_{t^-} = (y_{t-1}, y_{t-2}, \ldots, y_{t-\nu})'$$

and

$$Y_{t^+} = (y_t, y_{t+1}, \ldots, y_{t+\nu-1})'.$$

Then

---

* Some recommend dividing by N-t rather than N to preserve positive semi-definiteness of the covariance matrix. See Section 8.3 and van Zee [1981].

$$\mathcal{H}_\nu = E\{Y_{t^+} Y_{t^-}'\} = E\{Y_{o^+} Y_{o^-}'\}$$

where the second equality follows from the assumed weak stationarity of $\{y_t\}$.

The dimension of the state space model can be characterized in at least two equivalent ways. The identification example suggests that the order of the ARMA model is given by the smallest positive integer n such that rank $\mathcal{H}_n$ = rank $(\mathcal{H}_{n+i})$ for all $i \geq 1$. Alternatively, the conditional prediction example suggests that the order is the smallest positive integer n such that $y_{t+n|t-1}$ is linearly dependent on the predecessors $y_{t+i|t-1}$, i = 0, 1 ..., n-1, i.e., as the first row of $\mathcal{H}$ which is linearly dependent on the previous rows. These two are equivalent because $\mathcal{H}_\nu$ has rank n if and only if it has only n linearly independent rows. This can be seen explicitly as follows:

$$\mathcal{H}_\nu = E\{Y_{t^+} Y_{t^-}'\}$$

$$= E\begin{pmatrix} y_t \\ \vdots \\ y_{t+\nu-1} \end{pmatrix} [y_{t-1}' \cdots y_{t-\nu}']$$

$$= E\begin{pmatrix} y_{t|t-1} \\ \vdots \\ y_{t+\nu-1|t-1} \end{pmatrix} [y_{t-1}' \cdots y_{t-\nu}']$$

$$= E\begin{pmatrix} y_{t|t-1} Y_{t^-}' \\ \vdots \\ y_{t+\nu-1} Y_{t^-}' \end{pmatrix}.$$

We construct a state space model of the time series associated with this Hankel matrix by the singular value decomposition of this Hankel matrix. Because canonical correlation also obtains a singular value decomposition of "normalized" correlation matrix, this procedure, therefore, reminds us a closely related method of constructing models of time series by canonical correlation proposed and implemented by Akaike [1976].

We briefly describe the basic idea. Let x and y be two column vectors of dimension p and q respectively, where $p \le q$ to be definite and with the covariance matrix $\text{cov}\begin{pmatrix} x \\ y \end{pmatrix} = \begin{pmatrix} \Sigma_{11} & \Sigma_{12} \\ \Sigma_{21} & \Sigma_{22} \end{pmatrix}$. To be definite we assume that Σ_{11} and Σ_{22} are nonsingular and rank $\Sigma_{12} = r$. Let the singular value decomposition of $\Sigma_{11}^{-1/2}\Sigma_{12}\Sigma_{22}^{-1/2}$ be UPV'. Because the rank of $\Sigma_{11}^{-1/2}\Sigma_{12}\Sigma_{22}^{-1/2}$ is also r, the matrix P is of the form $P = [\text{diag}\,(\rho_1 \ldots \rho_r 0 \ldots 0),\ 0] : p{\times}q$. Define L and M to equal $U'\Sigma_{11}^{-1/2}$ and $V'\Sigma_{22}^{-1/2}$ respectively. Let $u = Lx$ and $v = My$. Then $\text{cov}\ u = U'U = I_p$, $\text{cov}\ v = V'V = I_q$ and $\overline{uv'} = U'\Sigma_{11}^{-1/2}\Sigma_{12}\Sigma_{22}^{-1/2}V = P$. The change of variables from (x, y) to (u, v) causes the covariance matrix of the transformed variables to have a special structure: $\text{cov}\begin{pmatrix} u \\ v \end{pmatrix} = \begin{pmatrix} I_p & P \\ P' & I_q \end{pmatrix}$. This covariance matrix structure shows that $P = E(uv')$ is the correlation matrix between two vectors u and v, each normalized to have unit variances. The components of the vector u are called the canonical variables of x, those of v of y. The column vectors of L' and M' are called canonical vectors. (Note the similarities with the definition of the principal components. There, a given vector x is represented as $\Gamma'x$ where the column vectors of Γ are the eigenvectors of the covariance matrix of x. Here two vectors are involved.) The canonical variables have unit variances. They have covariances (which are equal to the correlations because of unit variances and zero means)

$$\text{cov}\,(u_i, v_j) = \begin{cases} \rho_i, & i = j \quad 1 \le i, j \le r, \\ 0, & \text{otherwise.} \end{cases}$$

The positive square roots of the singular values of $\Sigma_{11}^{-1/2}\Sigma_{12}\Sigma_{22}^{-1/2}$, ρ_1, ..., ρ_r, are called the canonical correlations. Suppose that x and y are jointly normal. Then the conditional mean of y given x is $(\Sigma_{11}^{-1}\Sigma_{12})'x$ when Ex and Ey are both zero, i.e., the matrix $\Sigma_{11}^{-1}\Sigma_{12}$ is recognized as the matrix of regression coefficients in regressing y on x.

10 INNOVATION PROCESSES

This chapter constructs innovation models to reproduce second order properties of given time series. This construction phase completes the initial of the process of building dynamic models for vector-valued time series which was started in Section 9.1.

10.1 Orthogonal Projection

Minimization of $E(z-w'y)^2$ with respect to w is a well-known problem of constructing the least squares estimate of z, or projecting z orthogonally onto the subspace spanned by y where y is a mean-zero finite variance random vector. The components of w are the weights associated with individual data, i.e., components of y.

The best w is obtained as

(1) $$w = \Sigma_{yy}^{-1}\Sigma_{yz}$$

where

$$\Sigma_{yy} = E(yy'), \quad \text{and} \quad \Sigma_{yz} = E(yz').$$

The orthogonal projection of z onto the subspace spanned by y is thus expressible as

(2) $$\begin{aligned} w'y &= \Sigma'_{yz}\Sigma_{yy}^{-1} y \\ &= \Sigma_{zy}\Sigma_{yy}^{-1} y. \end{aligned}$$

We denote this by $\tilde{E}(z|y)$ and call it the wide sense conditional expectation or the best linear (least squares) prediction of z given the data vector y.*

---

* If z and y are jointly normally distributed, then the conditional probability density is explicitly calculated to verify that (2) equals $E(z|y)$. See Aoki [1967] for example.

When a new piece of data u is added to the existing data y, the weights (1), and the least squares estimate (2) are all altered in general. However, by extracting a component from the new data that is uncorrelated with the existing data we can calculate $\tilde{E}(z|y, u)$ quite easily. This should all be very familiar to the reader who knows the Gram-Schmidt orthogonalization method in nonlinear programing or from our discussion of the Cholesky decomposition. The component of u which is uncorrelated with y is simply that part of u that is orthogonal to the subspace spanned by y

$$e = u - \tilde{E}(u|y) = u - \overline{uy'}(\overline{yy'})^{-1}y. \tag{3}$$

We can verify that e and y are indeed uncorrelated by calculating $E(ey') = 0$.

The least squares estimate of z, given y and u is the same as the least squares estimate of z, given y and e. From (2)

$$\tilde{E}(z|y, e) = \overline{z(y'e')}\begin{pmatrix} \overline{yy'}^{-1} & 0 \\ 0 & \overline{ee'}^{-1} \end{pmatrix}\begin{pmatrix} y \\ e \end{pmatrix} = \overline{zy'}\ \overline{yy'}y + \overline{ze'}\ \overline{ee'}^{-1}e \tag{4}$$

i.e.,

$$\tilde{E}(z|y, e) = \tilde{E}(z|y) + \tilde{E}(z|e).$$

Here we take advantage of the uncorrelatedness of y and e:

$$\text{cov}\begin{pmatrix} y \\ e \end{pmatrix} = \begin{pmatrix} \overline{yy'} & 0 \\ 0 & \overline{ee'} \end{pmatrix}.$$

We can formally establish that $\tilde{E}(z|y, e) = \tilde{E}(z|y, u)$ by applying the coordinate transformation

$$\begin{pmatrix} y \\ u \end{pmatrix} = \begin{pmatrix} I & 0 \\ \overline{uy'}\ \overline{yy'}^{-1} & I \end{pmatrix}\begin{pmatrix} y \\ e \end{pmatrix}$$

$$\text{to } \tilde{E}(z|y,\ u) = \overline{z(y'u')}\,\overline{\begin{pmatrix} y \\ u \end{pmatrix}(y'u')}^{-1}\begin{pmatrix} y \\ u \end{pmatrix}.$$

The required modification to the best least squares prediction is achieved by merely calculating $\tilde{E}(z|e)$ as the correction term to the previous best prediction.

We call e constructed by (3) the innovation of the new data. Now introduce time index explicitly and number the data vector as $y_1, y_2, \ldots$ and construct the innovations thus:

$$e_1 = y_1$$

$$e_i = y_i - \tilde{E}(y_i|y_{i-1}, \ldots, y_1),\ i = 2, 3, \ldots .$$

Then

$$\tilde{E}(z|y_n, y_{n-1}, \ldots, y_1) = \Sigma_{i=1}^{n} \tilde{E}(z|e_i) \tag{5}$$

$$= \Sigma_{i=1}^{n} K_i \Sigma_i^{-1} e_i$$

where

$$K_i = \overline{ze_i'} \quad \text{and} \quad \Sigma_i = \overline{e_i e_i'}.$$

We see that the best estimate or prediction of z, given observations $y_1, \ldots, y_n$ from a time series $\{y_t\}$, is expressed by an MA(n) model or as a sum of n separate innovation terms: a finite data version of the Wold decomposition.

By construction these e's are uncorrelated. Let $D = \text{diag}\ (D_1, \ldots, D_n)$, and perform the Cholesky factorization of Σ_{yy}, where $y' = (y_1', \ldots, y_n')$, as

$$\Sigma_{yy} = LDL'$$

where L is a block lower triangular matrix which is the same as in the Gram-Schmidt orthogonalization of the y vector (see Aoki [1971; p.3]). Then we can write

$$\begin{pmatrix} e_1 \\ \vdots \\ e_n \end{pmatrix} = L^{-1}y.$$

It is easy to see that $\{e_i\}$ are serially uncorrelated:

$$E\begin{pmatrix} e_1 \\ \vdots \\ e_n \end{pmatrix}[e_1' \ \dots \ e_n'] = L^{-1}Eyy'L'^{-1}$$

$$= L^{-1}\Sigma_{yy}L'^{-1}$$

$$= L^{-1}LDL'L'^{-1}$$

$$= D.$$

Because L is lower triangular, the operation can be reversed to generate y_i from $e_1, \dots, e_i$ above:

$$y = L\begin{pmatrix} e_1 \\ \vdots \\ e_n \end{pmatrix}.$$

The sequence $\{e_i\}$ and $\{y_i\}$ are therefore said to be causally equivalent because they span the same subspace (σ-field).

10.2 Kalman Filters

Suppose a Markov model for the time series $\{y_t\}$ is known to be given by

(1)
$$\begin{cases} z_{t+1} = Az_t + u_t \\ y_t = Cz_t + v_t, \end{cases}$$

where the noises are mean-zero and serially uncorrelated with covariance matrices

$$\operatorname{cov}\begin{pmatrix} u_t \\ v_t \end{pmatrix} = \begin{pmatrix} Q_t & N_t \\ N_t' & R_t \end{pmatrix}.$$

Start the Kalman filter for this model at time 1. It starts calculating the (wide sense) conditional mean of z_{t+1} given the data $\{y_1, \dots, y_t\}$, $t \geq 1$.

Denote the data by y_1^t for shorthand notation. In the previous section we have established that (see (1.5))

$$\begin{aligned} z_{t+1|t} &= \tilde{E}(z_{t+1}|y_1^t) \\ &= \tilde{E}(z_{t+1}|y_1^{t-1}, e_t) \\ &= \tilde{E}(z_{t+1}|y_1^{t-1}) + \tilde{E}(z_{t+1}|e_t) \\ &= z_{t+1|t-1} + K_t\Sigma_t^{-1} e_t \end{aligned}$$

where $e_t = y_t - \tilde{E}(y_t|y_1^{t-1})$ because e_t is uncorrelated with y_1^{t-1} by construction, and where

$$K_t = E(z_{t+1}e_t'),\text{ and } \Sigma_t = E(e_t e_t').$$

So far the Markov structure of (1) has not been utilized. We use the Markov property of the model when we relate z_{t+1} to z_t by $z_{t+1} = Az_t + u_t$, and write $z_{t+1|t-1}$ as

$$\begin{aligned} z_{t+1|t-1} &= Az_{t|t-1} + u_{t|t-1} \\ &= Az_{t|t-1}, \end{aligned}$$

since $E(y_1^{t-1}u_t') = 0$.

The Kalman filter for (1) can thus be written as

$$z_{t+1|t} = Az_{t|t-1} + K_t\Sigma_t^{-1} e_t \tag{2}$$

where

$$\begin{aligned} e_t &= y_t - y_{t|t-1} \\ &= y_t - Cz_{t|t-1} \end{aligned}$$

because $\tilde{E}(v_t|y_1^{t-1})$ is also zero.

The sequence $\{e_t\}$ has been shown to be serially uncorrelated in the previous section. It is the innovation sequence associated with (1).*

We need various covariances associated with (1) and (2) to proceed further. First, define the prediction error covariance of the Kalman filter by P_t

(3) $$P_t = E(z_t - z_{t|t-1})(z_t - z_{t|t-1})'$$
$$= \Pi_t - Z_t \geq 0,$$

where Π_t and Z_t are covariances of z_t and $z_{t|t-1}$ respectively;

$$\Pi_t = E(z_t z_t'),$$

and

$$Z_t = E(z_{t|t-1} z'_{t|t-1}).$$

The recursion (2) shows that the dynamics for Z's are

(4) $$Z_{t+1} = AZ_t A' + K_t \Sigma_t^{-1} K_t'$$

because $E(z_{t|t-1} e_t') = 0$.

The Markov model (1) yields the recurson of Π_t

(5) $$\Pi_{t+1} = A\Pi_t A' + Q_t.$$

The filter gain K_t becomes, after z_{t+1} is substituted out by (1) and e_t by (2)

(6) $$K_t = E(z_{t+1} e_t')$$
$$= E(Az_t + u_t)\{C(z_t - z_{t|t-1}) + v_t\}',$$
$$= AP_t C' + N_t$$

where we use the fact that $z_t - z_{t|t-1}$ is uncorrelated with, i.e., orthogonal to, $z_{t|t-1}$. Similarly we can write

(7) $$\Sigma_t = E(e_t e_t')$$
$$= CP_t C' + R_t.$$

---

* Unless z_0, u's and v's are all Gaussian in (1), e's are, strickly speaking, pseudo-innovations.

Advancing t by one in (3) and taking the difference of the two recursion relations (4) and (5), we deduce the recursion formula for P_t*

(8) $$P_{t+1} = \Pi_{t+1} - Z_{t+1}$$
$$= AP_tA' + Q_t - K_t\Sigma_t^{-1}K_t'.$$

The covariance matrices of $\{y_t\}$ are related to these matrices by

(9) $$Y_t = Ey_ty_t'$$
$$= C\Pi_tC' + R_t$$

and by expressing y_t in terms of z_s and the noises and using (1), we calculate

(10) $$\Lambda_{t,s} = E(y_ty_s')$$
$$= CA^{t-s-1}(A\Pi_sC' + N_s), \quad t > s.$$

The derivations so far show that Kalman filters can deal with nonstationary noises (and time-varying dynamics although we have not done so explicitly). For our analysis of time series, however, we now assume that noises are wide-sense

---

* Equation (8) can be directly obtained as follows. From (1) and (2) $e_t = C(z_t-z_{t|t-1}) + v_t$. Hence $z_{t+1} - z_{t+1|t} = A(z_t-z_{t|t-1}) + u_t - K_t\Sigma_t^{-1}e_t = (A-K_t\Sigma_t^{-1}C)\cdot(z_t-z_{t|t-1}) + u_t - K_t\Sigma_t^{-1}v_t$. Taking the covariance of this, we note that

$$P_{t+1} = (A-K_t\Sigma_t^{-1}C)P_t(A-K_t\Sigma_t^{-1}C)' + Q_t$$
$$- N_t\Sigma_t^{-1}K_t' - K_t\Sigma_t^{-1}N_t' + K_t\Sigma_t^{-1}R_t\Sigma_t^{-1}K_t'.$$

Use the identity (7) $\Sigma_t = CP_tC' + R_t$ to write the quadratic term in K_t as $K_t\Sigma_t^{-1}K_t'$. Collect the terms linear in K_t as $-K_t\Sigma_t^{-1}CP_tA - K_t\Sigma_t^{-1}N_t' = -K_t\Sigma_t^{-1}(AP_tC + N_t)' = -K_t\Sigma_t^{-1}K_t'$, where (5) is used. P_{t+1} can be written as (8) because

$$P_{t+1} = AP_tA' + Q_t - 2K_t\Sigma_t^{-1}K_t' + K_t\Sigma_t^{-1}K_t'$$
$$= AP_tA' + Q_t - K_t\Sigma_t^{-1}K_t'.$$

stationary, dropping the time subscript from noise covariances, and assume A and C are constant matrices.

Suppose now we examine the consequence of starting the Kalman filter off from some remote past. First, let the subscript t be replaced by $t-t_0$ to indicate that the filter is started at time t_0. If the noise sequences are wide sense stationary and A is asymptotically stable, then letting t_0 recede into the past has the same effect on the recursions for $\{\Pi_t\}$ as letting t approach infinity. Hence, denoting the steady state (limiting) value of Π_t by Π as $t \to \infty$, (5) shows that it satisfies an algebraic matrix equation*

$$\Pi = A\Pi A' + Q,$$

and from (9) $\{y_t\}$ becomes stationary with the covariance matrices

(11) $$Y_t = Y_0 = C\Pi C' + R, \qquad t \geq t_0,$$

while (10) shows that

$$E(y_t y_s') = \Lambda_{t-s} = CA^{t-s-1}(A\Pi C' + N), \qquad t > s.$$

In symbol Λ_0 was used in Chapter 9 Y_0.

Similarly (8) becomes an algebraic equation for the limiting matrix of P_t

$$P = APA' + Q - K\Sigma^{-1}K',$$

where (6) shows that

(12) $$K = APC' + N$$

and the limit of the innovation covariance matrix

(13) $$\Sigma = CPC' + R,$$

follows from (7). P denotes the limiting value of P_t as $t \to \infty$ or equivalently as $t_0 \to -\infty$. Furthermore $P = \Pi - Z$ also holds, and (4) tells us that $Z = AZA' + K\Sigma^{-1}K'$.

---

* For the existence of the limit see Lyapunov Theorem in Appendix A.6.

From (11) and (13), the matrix K is alternatively given by

$$\Sigma = Y_0 - CZC'.$$

Letting $t_0 \to -\infty$, (3) shows that $\Pi = Z + P \geqslant Z$, hence $\pi_* = Z$, i.e., the covariance associated with the Kalman filter (2) attains the minimum among all π's, i.e., among the covariance matrix of the state vector z_t. From (4), π_* satisfies $\pi_* = A\pi_* A' + K\Sigma^{-1}K'$. Recalling Section 9.1, we recognize that, given A, C and M, π_* may be iteratively calculated as follows (see also (4.1) below):

$$\begin{cases} \Omega_0 = 0, \\ \Omega_{n+1} = A\Omega_n A' + (M - A\Omega_n C')(Y_0 - C\Omega_n C')^{-1}(M - A\Omega_n C')', \\ \pi_* = \lim_{n\to\infty} \Omega_n. \end{cases}$$

Earlier in Section 9.1, phase one of the process of constructing dynamic models has been explained. In phase two, we estimate noise covariances assuming that A, C and M are known by

$$R = \Lambda_0 - C\pi_* C'$$

$$Q = \pi_* - A\pi_* A'$$

and

$$N = M - A\pi_* C'.$$

10.3 Innovation Model

The previous section is based on model (2.1) which is the same as model (1) of Section 9.1. We now construct the innovation model, model (I) below, for a time series $\{y_t\}$. The innovation model is important bacuase it is causally invertible and can immediately be computed from the Kalman filter. Generate matrix sequences $\{T_t\}$, $\{\Omega_t\}$ and $\{L_t\}$, corresponding to π, Σ and K of the previous section, by

$$T_{t+1} = AT_tA' + (AT_tC'-M_t)(Y_t-CT_tC')^{-1}(AT_tC'-M_t)',$$

$$T_0 = 0,$$

$$\Omega_t = Y_t - CT_tC',$$

and

$$L_t = M_t - AT_tC',$$

where we assume that Ω_t is positive definite. (Otherwise use its pseudo-inverse.)

Note that the recursion for T_t can be written as

$$T_{t+1} = AT_tA' + L_t\Omega_t^{-1}L_t'$$

which is exactly the same as (4) if we identify T_t with Z_t, L_t with K_t and Ω_t with Σ_t. Similarly, comparison of these recursions with those of (6), (7) and (9) reveals that Ω_t corresponds to Σ_t, and L_t to K_t, because (6) shows that $K_t = -AZ_tC' + M_t$ where from (10) $M_t = A\Pi_tC' + N_t$; (7) and (9) show that $\Sigma_t = CP_tC' + R_t = C\Pi_tC' + R_t - CZ_tC' = Y_t - CZ_tC'$.

We define the innovation model for $\{y_t\}$ to be

$$\text{(I)} \qquad \begin{aligned} \zeta_{t+1} &= A\zeta_t + L_t\Omega_t^{-1}\varepsilon_t, \qquad \zeta_0 = 0 \\ w_t &= C\zeta_t + \varepsilon_t \end{aligned}$$

where ζ_t is the state vector of the model, w_t is its output vector which reproduces the second order properties of the y's, and

$$E\varepsilon_t = 0, \quad E\varepsilon_t\varepsilon_s' = \Omega_t\delta_{t,s}.$$

Equation (2.2) is the Kalman filter of the model (2.1). Compare (2.2) with (I) to note that $\{w_t\}$ is a realization of $\{y_t\}$ and $E(\zeta_t\zeta_t') = T_t$. From the zero initial conditions for ζ_0 and T_0, we also note that the covariance of $\{w\}$ coincide with $\{\Lambda_t\}$ of (1): noting that $L_0 = M_0$ because of $T_0 = 0$,

$$\begin{aligned} E(w_tw_0') &= CA^{t-1}L_0 \\ &= CA^{t-1}M_0 \\ &= \Lambda_t. \end{aligned}$$

Conversely, a model with the state vector s_t and output vector y_t

(1) $$\begin{cases} s_{t+1} = As_t + L_t\Omega_t^{-1}\varepsilon_t, & s_0 = 0 \\ y_t = Cs_t + \varepsilon_t \end{cases}$$

with ε_t as in (I) is the innovation model for $\{y_t\}$ if Ω_t is nonsingular.

We verify this claim by showing that the gain of the Kalman filter associated with (1) is exactly $L_t\Omega_t^{-1}$ and that the innovation covariance is exactly Ω_t. First, construct its Kalman filter:

(2) $$s_{t+1|t} = As_{t|t-1} + \Gamma_t(y_t-Cs_{t|t-1}), \qquad s_{0|-1} = 0,$$

where the Kalman filter gain is

$$\Gamma_t = \overline{s_{t+1}\varepsilon_t'}\,\Omega_t^{-1}$$

$$= L_t\Omega_t^{-1}.$$

Because (1) and (2) have the same initial condition and the same dynamic equation, we conclude that $s_t = s_{t|t-1}$. This can be seen in another way: From (1) $s_{t+1} = As_t + L_t\Omega_t^{-1}(y_t-Cs_t)$, $s_0 = 0$. This shows that s_t is exactly computed from y , $\tau < t$, i.e., $E(s_t|y_0^{t-1}) = s_t$. Hence (1) can be written as

$$s_{t+1|t} = As_{t|t-1} + L_t\Omega_t^{-1}(y_t-Cs_{t|t-1}).$$

Once $s_t = s_{t|t-1}$ is established in (1), the innovation covariance equals $\text{cov}\,(y_t-Cs_{t|t-1}) = \text{cov}\,\varepsilon_t = \Omega_t$.

Using the steady state (limit) values of K and Σ, the innovation model corresponding to (2.1) with initial time in the infinitely remote past is obtained:

(3) $$\begin{cases} s_{t+1} = As_t + K\Sigma^{-1}e_t \\ w_t = Cs_t + e_t \end{cases}$$

where

$$E(e_t) = 0, \quad E(e_te_t') = \Sigma.$$

For this model $E(s_t s_t') = Z = AZA' + K\Sigma^{-1}K'$, the steady state version of (2.4) holds. The dynamics are stable if and only if the matrix $A - K\Sigma^{-1}C$, which results after substituting e_t out by $w_t - Cs_t$, has all its eigenvalues inside the unit disc. This follows if $(A, K\Sigma^{-1})$ is stabilizable and (A, C) is observable. See Aoki [1967, 1976].

Causal Invertibility

Eliminate ε_t from the innovation model (I) to rewrite $\{\zeta_t\}$ as

$$\zeta_{t+1} = A\zeta_t + \Gamma_t(w_t - C\zeta_t), \qquad \zeta_0 = 0$$

where

$$\Gamma_t = L_t\Omega_t^{-1}.$$

Because of the zero initial condition $\zeta_0 = 0$, ζ_t is completely determined by w_0, ..., w_{t-1}. These, in turn, show that ε_t is determined by w_t and the past w's.

The model (3) is causally invertible. To see this, suppose that the Kalman filter for (3) is turned on at $t = t_0$. Then

$$Z = E(s_t s_t') = E(s_{t|t-1} s_{t|t-1}') + P_{t-t_0}$$

$$= Z_{t-t_0} + P_{t-t_0}.$$

The limiting operation $t_0 \to -\infty$ causes $Z_{t-t_0} \to Z$, hence $P_{t-t_0} \to 0$ or $s_{t|t-1} \xrightarrow{m.s.} s_t$, i.e., in the mean-square sense. From (3)

$$s_{t+1} = (A - K\Sigma^{-1}C)s_t + K\Sigma^{-1}w_t.$$

A Markov model (2.1) with a stable A matrix, i.e., where the eigenvalues of A are all less than one in magnitude is called causally invertible if one can construct sequences $u_t(t_0)$, $v_t(t_0)$ from y_{t_0}, y_{t_0+1}, ..., y_t, such that $u_t(t_0) \xrightarrow{m.s.} u_t$ and $v_t(t_0) \xrightarrow{m.s.} v_t$ as $t_0 \to -\infty$.

Define $v_t(t_0) = w_t - C\zeta_{t|t-1}$ where $\zeta_{t|t-1}$ is being computed from w_s, $s = t_0$, ..., $t-1$. Then $v_t(t_0) \to w_t - C\zeta_t$ as $t_0 \to -\infty$. Conversely, suppose (2.1) is invertible. From (2.1) for any $t > t_1$ we can write $z_t = z_t^1 + z_t^2$ where $z_t^1 = A^{t-t_1} z_{t_1}$ and $z_t^2 = \Sigma_{t_1}^{t-1} A^{t-1-j} Be_j$, where $B = K\Sigma^{-1}$. Given $\varepsilon > 0$ choose t_1 so that $E(z_t^1 z_t^{1\prime}) < \varepsilon I$. This is possible because the magnitudes of eigenvalues of A are less than one by assumption. Choose t_0 such that $y_{t_0}, y_{t_0+1}, \ldots, y_t$ define an estimate of z_t^2, with an error covariance less than εI. Let this estimate of z_t be denoted by $\hat{z}_t$.

Then

$$E\| z_t - \hat{z}_t \|^2 = E\| z_t^1 + z_t^2 - \hat{z}_t \|^2$$

$$\leq 2E\| z_t^1 \|^2 + 2E\| z_t^2 - \hat{z}_t \|^2 \leq 4n\varepsilon$$

where $n = \dim \zeta_t$.

Now $\operatorname{cov}(z_t) = \operatorname{cov}(z_t - z_{t|t-1}) + Z_t(t_0)$. As $t_0 \to -\infty$, $\operatorname{cov}(z_t - z_{t|t-1}) \to 0$ hence $E(z_t z_t') = \Pi = Z$. Also from this the following equalities hold:

$$E(y_t y_0') = CA^{t-1}M,$$

$$M = A\Pi C' + N,$$

$$M = AZC' + K.$$

From $\Pi = Z$ follows $N = K$, and $E(y_0 y_0') = \Lambda_0 = C\Pi C' + R$ from the Markov model, and $\Lambda_0 = C\Pi C' + \Sigma$ from the innovation model. Since $\Pi = Z$, $R = \Sigma$. From (2.1) $\Pi - A\Pi A' = Q$ and from the innovation model $\Pi = Z$, $Z - AZA' = K\Sigma^{-1}K'$.

10.4 Output Statistics Kalman Filter

The previous sections use noise covariance information in their Kalman filter calculation. Son and Anderson [1973] give alternative expressions without noise covariances.

Write K_t as*

$$K_t = E(z_{t+1}e_t')$$

$$= Ez_{t+1}(y_t' - z_{t|t-1}'C')$$

$$= Ez_{t+1}y_t' - AZ_tC'$$

and Σ_t as

$$\Sigma_t = Ey_ty_t' - CZ_tC'$$

$$= Y_t - CZ_tC'.$$

Substitute P_t out by (2.3) in (2.6) to express the gain matrix as

$$K_t = A(\Pi_t - Z_t)C' + N_t$$

$$= A\Pi_tC' + N_t - AZ_tC'$$

$$= M_t - AZ_tC'$$

where

$$M_t = A\Pi_tC' + N_t$$

and recognize that M_t appears in

$$\Lambda_{t-s} = E(y_ty_s') = CA^{t-s-1}M_s, \qquad t > s.$$

The recursion for Z_t given by (2.4) then can be rewritten as

(1) $$Z_{t+1} = AZ_tA' + (M_t - AZ_tC')(Y_t - CZ_tC')^{-1}(M_t - AZ_tC')'.$$

where

$$Y_t = E(y_ty_t'), \qquad Z_0 = 0.$$

10.5 Spectral Factorization

Because $\Lambda_t = E(y_ty_0') = CA^{t-1}M$, the spectrum of $\{y_t\}$ is given by

$$S(z) = \Lambda_0 + C(zI-A)^{-1}M + M'(z^{-1}I-A')^{-1}C'$$

---

* Solo [1983] claims that $E(z_{t+1}y_t')$ can be obtained in a model fitting exercise.

$$= H(z) + H'(z^{-1})$$

where

$$H(z) = \Lambda_0/2 + C(zI-A)^{-1}M.$$

Define W(z) as the transfer function of the innovation model

$$W(z) = I + C(zI-A)^{-1}K\Sigma^{-1}$$

where, with the initial condition specified as the infinitely remote past, the following relations hold:

$$\Lambda_0 = C\Pi C' + R = \Sigma + CZC', \qquad = CA^{\ell-1}M,$$

$$Z = AZA' + K\Sigma^{-1}K', \qquad M = A\Pi C' + N,$$

$$\Pi = A\Pi A' + Q,$$

$$K = M - AZC'.$$

Then the spectrum S(z) can be factored as*

$$S(z) = W(z)\Sigma W'(z^{-1}).$$

Note that the matrix W(z) can be constructed from

---

* First, form the product

$$\{I + C(zI-A)^{-1}K\Sigma^{-1}\}\Sigma\{I+\Sigma^{-1}K'(z^{-1}I-A')^{-1}C'\}$$
$$= \Sigma + C(zI-A)^{-1}K + K'(z^{-1}I-A')^{-1}C'$$
$$+ C(zI-A)^{-1}K\Sigma^{-1}K'(z^{-1}I-A')^{-1}C'.$$

Substitute Λ_0 - CZC' for Σ and M - AZC' for K then collect terms as follows:

$$W(z)\Sigma W'(z) = \Lambda_0 + C(zI-A)^{-1}M + M'(z^{-1}I-A')^{-1}C$$
$$+ C(zI-A)^{-1}D(z^{-1}I-A')^{-1}C'$$

where noting that $K\Sigma^{-1}K'$ equals Z - AZA', D is identically equals zero where

$$D = Z - AZA' - (zI-A)Z(z^{-1}I-A')$$
$$- AZ(z^{-1}I-A') - (zI-A)ZA' = 0.$$

This establishes the factorization.

$$\begin{pmatrix} Z-AZA' & M-AZC' \\ (M-AZC')' & \Lambda_0-CZC' \end{pmatrix} = \begin{pmatrix} K\Sigma^{-1}K' & K \\ K' & \Sigma \end{pmatrix}$$

$$= \begin{pmatrix} K\Sigma^{-1} \\ I \end{pmatrix} \Sigma [\Sigma^{-1}K' \quad I].$$

When Z equals Π, the above matrix can be written as

$$\begin{pmatrix} \Pi-A\Pi A' & M-A\Pi C' \\ (M-A\Pi C')' & \Lambda_0-C\Pi C' \end{pmatrix} = \begin{pmatrix} Q & N \\ N' & R \end{pmatrix}$$

i.e.,

$$\begin{pmatrix} Q & N \\ N' & R \end{pmatrix} = \begin{pmatrix} K\Sigma^{-1} \\ I \end{pmatrix} \Sigma [\Sigma^{-1}K', I].$$

11 TIME SERIES FROM INTERTEMPORAL OPTIMIZATION

Economic time series are generated as economic agents engage in intertemporal optimization. Although time is an extra complicating factor, dynamic optimization, i.e., optimization over time arises for the same reason that static optimization (i.e., linear and nonlinear programming) problems arise in economics: Trade-offs must be made in allocating scarce resources; the only difference being that the trade-offs over time also must be made because dynamics constrains choice sets effectively over time. Economic time series are usually nonstationary because circumstances facing optimizing economic agents change with time and do not remain the same. Time series are also nonlinear because the dynamic structure generating data are mostly nonlinear. We are thus faced with nonstationary and nonlinear stochastic processes.

Intertemporal optimization of dynamic systems can best be approached using Markovian or state-space representation of dynamic structure. This point of view is inherent in dynamic programming and has been vigorously pursued in the systems literature. Some examples to be introduced presently illustrate how state-space representation may naturally arise in economic intertemporal optimization problems.

It should come as no surprise that theory of dynamic optimization is best developed for linear dynamic systems. Furthermore, optimization of linear dynamic systems with quadratic performance indices can be developed in an elementary and a self-contained way without elaborate theory. Dynamic programming, when it leads to closed form solutions, is most effective and conceptually straightforward. Linear dynamic systems with quadratic separable cost or performance indices constitute an important class of intertemporal problems which yield explicit closed form optimization rules by dynamic programming. For this reason, we begin our discussion of dynamic optimization with linear dynamic systems with quadratic costs. Optimization of some nonlinear dynamic systems with not-necessarily quadratic performance indices may be iteratively approximated by solving sequence of optimization problems for linear dynamic systems with quadratic costs (Aoki [1962]). This further motivates our study of linear dynamic system optimizations with quadratic costs.

When optimization problems with nonquadratic costs or nonlinear dynamics do not yield explicit analytical solutions by dynamic programming, we have no generally valid analytical tools for dealing with them. We must resort to procedures to approximate nonstationary, nonlinear phenomena by locally stationary and locally linear ones. We can proceed in at least two ways. In one approach nonlinear dynamic systems are studied as deviation from some reference paths as we discuss in Chapter 6, i.e., decision or choice variables that are normally chosen to guide nonlinear dynamic systems along some reference paths are assumed known. (In the language of control theory, reference decision variables cause the nonlinear system to "track" or follow the reference time path.) We then focus on their deviational effects as the decision variables respond to deviations in exogenous variables causing the model to go off the reference paths. In this way, deviation of the actual time path from the reference paths are described by (variational) linear dynamic equations. (See Aoki [1976; pp.59-62] or Aoki [1981; Chapter 2] for more detailed description of the procedure. Examples in macroeconomics are found in Aoki [1976; pp.66-68, 239-243] and many places in Aoki [1981].) In econometrics linear (time series) models are often specified for variables that are logarithms of "more basic" variables, yielding so-called log-linear models. These models may be interpreted to arise in the way we described above as deviational or variational models. These models are then converted to state space form to apply a body of well-developed theory for dynamic optimization in state space form.

In the other way we do not explicitly approximate nonlinear problems, but rather directly work with first (and second) order necessary conditions for optimality. Necessary conditions for optimality rarely yield closed-form analytic solutions to optimization problems. Optimization problems are usually too complicated to permit explicit analytic solutions. Necessary conditions are more frequently used to characterize optimal solutions, to narrow a class of possible solutions over which search for optimal solutions are conducted. This is very well understood in the engineering literature. In economics, however, this seems to have been brought to the attention of the profession by Hall [1978].

Even when explicit closed form solutions are not available, first and second order optimality conditions are often useful in characterizing optimal solutions or reducing the class of solutions from which optimal ones are to be chosen. Following Hall [1978] a number of recent investigators has employed this approach effectively. Pontryagin's maximum principle is the most systematic way to derive such first and second order conditions. We quote one version in Appendix, which is based on Canon et al. [1970] for discrete time dynamics. For continuous time version, see Lee and Markus [1967], Flemming and Rischel [1975] or Kamien and Schwarz [1981].

11.1 Example: Dynamic Resource Allocation Problem*

We use a simplified model of Long and Plosser [1983] to illustrate how

---

* The model discussed in this section is a simplified version of the one in Long and Plosser [1983].

economic time series are generated as agents engage in dynamic, i.e., intertemporal optimization. As Long and Plosser mention, this model so allows the maximizing consumer with sufficient intertemporal and intratemporal substitution opportunities (i.e., among consumption goods and work vs. leisure) that he chooses to spread effects of unanticipated output shocks through time and across commodities. Thus, the output time series of various commodities can show both persistent fluctuations and comovements. This example captures one way that business cycles may result from such optimizaing behavior. This example serves yet another useful purpose because the concept of state introduced in Chapter 2 naturally arises in formulating the intertemporal optimization problem as a dynamic programming functional equation.

Consider a dynamic allocation decision problem in which an infinitely lived individual allocates his time between leisure and work and the available output between consumption and input for future production. First we discuss a deterministic version, then a stochastic version. The former is used to introduce and illustrate the dynamic programing procedure for formulating such intertemporal, i.e., sequential decision problems, in particular Bellman's principle of optimality. The latter is used to amplify on the notion of "state" of a dynamic system.

There are two activities producing 2 goods, each of which is to be consumed and also be used as inputs. In its deterministic version the problem is to maximize the present value at time t of the discounted sum of utilities given by

(1) $$U_t = \Sigma_{\tau=t}^{\infty} \beta^{\tau-t} u(C_\tau, Z_\tau), \qquad 0 < \beta < 1$$

where

$$u(C_\tau, Z_\tau) = \theta_0 \ln Z_\tau + \theta_1 \ln C_{1\tau} + \theta_2 \ln C_{2\tau} \qquad \theta_0 > 0, \quad \theta_i \geq 0, \; i = 1, 2,$$

Subject to the next three constraints:

(2) $$C_{jt} + X_{1jt} + X_{2jt} \leq Y_{jt}, \qquad j = 1, 2$$

(3) $$Y_{it+1} = L_{it}^{b_i} \prod_{j=1}^{2} X_{ijt}^{a_{ij}}$$

where

$$b_i + \Sigma_{j=1}^{2} a_{ij} = 1, \qquad i = 1, 2$$

and

(4) $$Z_t^1 + L_{1t} + L_{2t} = H.$$

The log-linear utility function is used to yield an analytically closed form solution. The leisure time is denoted by Z_t. In (2), X_{ijt} denotes the amount of good j allocated as input to produce good i. The time devoted to producing good i is denoted by L_{it} in (3). Equation (3) is the Cobb-Douglas production function. The parameters θ's, b's and a_{ij}'s express the individuals' preferences, and production technologies respectively and do not change. They are the structural parameters.

Since H remains constant, the knowledge of $Y_t = (Y_{1t}, Y_{2t})$ at time t completely specifies the maximum attained by U_t. For this reason we call Y_t the state vector of the problem. The constrained maximum of U_t is called the optimal value V_t. Since it depends only on Y_t we write it as

$$V(Y_t) = \max\{U_t \text{ subject to (2)} \sim \text{(4)}\}.$$

Note that U_t is maximized with respect to all current and future decision variables. The current allocation decision variables are L_{it}, C_{it}, X_{ijt}, i, j = 1, 2. Given the current decision, the immediate or period t return is $u(C_t, Z_t)$. The state is transformed into Y_{t+1} and the problem starts all over again, i.e., the problem of choosing $L_{i\tau}$, $C_{i\tau}$, $X_{ij\tau}$ for $\tau \geq t+1$ has the same structure as the decision problem facing the individual at time t. Given that optimal sequence of decisions are made form t+1 on, the maximum value is $V(Y_{t+1})$. Discounting the value from future optimal sequence of decisions the decision at t must, therefore, maximize the discounted sum $u(C_t, Z_t) + \beta V(Y_{t+1})$, i.e.,

$$V(Y_t) = \max_{d_t}\{u(C_t, Z_t) + \beta V(Y_{t+1})\} \tag{5}$$

where d_t stands for all current decision variables. Equation (5) thus stands for a sequence of nested decisions

$$V(Y_t) = \max_{d_t}\{u(C_t, Z_t) + \max_{d_{t+1}}\{u(C_{t+1}, Z_{t+1}) + \max_{d_{t+2}}\{u(C_{t+2}, Z_{t+2}) + \ldots\}\ldots\}.$$

If a sequence of decisions $\{d_t, d_{t+1}, d_{t+2}, \ldots\}$ is optimal, then the subsequence of decisions covering decisions from time t+1 on $\{d_{t+1}, d_{t+2}, \ldots\}$ must be optimal from time t+1 on. This is an illustration of Bellman's principle of optimality.

Equation (5) is a functional equation that $V(\cdot)$ must satisfy. In general this equation does not admits a closed form solution if a general $u(\cdot,\cdot)$ and a general production technology are employed. Our choice of the log-linear utility function and the Cobb-Douglas production function allows a closed form solution.*

Try

$$V(Y_t) = \gamma_1 \ln Y_{1t} + \gamma_2 \ln Y_{2t} + v_t. \tag{6}$$

Substituting this into the right hand side of (5) we note

$$\theta_0 \ln Z_t + \theta_1 \ln C_{1t} + \theta_2 \ln C_{2t} + \beta\{\gamma_1 \ln Y_{1t+1} + \gamma_2 \ln Y_{2t+1} + v_{t+1}\}.$$

After (3) is substituted into Y_{it+1}, $i = 1, 2$, maximizing the above is a static optimization problem solved by techniques of nonlinear programing. The first order conditions for optimality (these conditions are also sufficient for this problem) are:

$$Z = \theta_0/\lambda$$

$$L_{it} = \beta\gamma_i b_i/\lambda$$

---

* Another class of problem specifications allowing for closed form solutions are linear dynamics and quadratic objective functions.

$$C_{it} = \theta_i/\mu_i$$

$$X_{ijt} = \beta\gamma_i a_{ij}/\mu_j$$

where

$$\gamma_j = \theta_j + \beta\sum_i \gamma_i a_{ij}, \; j = 1, 2$$

where λ and μ_i are the Lagrange multipliers associated with (4) and (2) respectively. (We note that the inequality (2) is always binding for our problem, i.e., the inequality is replaced with the equality.)

Determine λ and μ_i from (2) and (4) as

$$\lambda = (\theta_0 + \beta\Sigma\gamma_i b_i)/H$$

and

$$\mu_j = \gamma_j/Y_{jt}.$$

Hence the optimal decisions are given by*

$$\begin{aligned} C^*_{it} &= (\theta_i/\gamma_i)Y_{it}, \\ X^*_{ijt} &= (\beta\gamma_i a_{ij}/\gamma_j)Y_{it}, \\ L^*_{it} &= H\beta\gamma_i b_i/(\theta_0 + \beta\Sigma\gamma_i b_i), \\ Z^* &= H\theta_0/(\theta_0 + \beta\Sigma\gamma_i b_i). \end{aligned} \tag{7}$$

The constant term in (6) evolve according to

$$v_t = \beta v_{t+1} + w$$

where

$$w = \theta_0 \ell n(\theta_0/\lambda) + \Sigma\theta_i \ell n(\theta_i/\gamma_i) + \beta\Sigma_i\gamma_i\{b_i \ell n(\beta\gamma b_i/\lambda) + \sum_j a_{ij}\ell n(\beta\gamma_i a_{ij}/\gamma_j)\}$$

$$= \theta_0 \ell n\theta_0 - \lambda H\ell n\lambda + \Sigma\theta_j \ell n\theta_j + \beta\ell n\beta\Sigma\gamma_i - (1 - \beta)\Sigma\gamma_i \ell n\gamma_i + \beta\Sigma\gamma_i\{b_i \ell nb_i + \sum_j a_{ij}\ell na_{ij}\}.$$

The transversality condition to ensure finite optimal value is

$$\lim_{\tau\to 0} \beta^\tau v_{t+\tau} = 0 \quad \text{for all } t \geq 0.$$

Then

$$v_t = w/(1 - \beta).$$

---

* In (7), γ's are derived parameters of the optimal decision rules.

Substituting (7) into (3), the optimal outputs are governed by

(8) $$y_{t+1} = Ay_t + n$$

where

$$A = (a_{ij}),\ y_t = \begin{pmatrix} y_{1t} \\ y_{2t} \end{pmatrix} \text{ where } y_{it} = \ell n Y_{it}$$

and

$$n_i = b_i \ell n(\beta\gamma_i b_i/\lambda) + \sum_j a_{ij} \ell n(\beta\gamma_i a_{ij}/\gamma_j).$$

We introduce stochastic elements by a random production or technological disturbance

$$Y_{it+1} = \lambda_{it+1}(L_{it}^{b_i} \Pi_j X_{ijt}^{a_{ij}})$$

where $\lambda_{t+1} = (\lambda_{1t+1}, \lambda_{2t+1})$ is assumed to be a Markovian process, i.e., the distribution function $F(\lambda_{t+1}|\lambda_t, \lambda_{t-1}, \ldots)$ equals $F(\lambda_{t+1}|\lambda_t)$. We assume that the value of λ_{t+1} becomes known at time t+1. The notion of state must now be enlarged to include λ_t because Y_t and λ_t now completely determine future evolution of Y's and λ's. Also we now maximaize the expected discounted streams of utilities.

Equation (5) is replaced with

$$V(S_t) = \max_{d_t}\{u(C_t, Z_t) + E(V(S_{t+1})|S_t)\}$$

where

$$S_t = (Y_t, \lambda_t).$$

Equation (6) changes into

$$V(S_t) = \gamma_1 \ell n Y_{1t} + \gamma_2 \ell n Y_{2t} + N(\lambda_t) + v_t$$

where

$$N(\lambda_t) = \beta E\{\Sigma_i \gamma_i \ell n \lambda_{it+1} + N(\lambda_{t+1})|S_t\}.$$

With these changes optimal decisions given by (7) remain valid. The dynamics for y_t now are stochastic, however, given by

(9) $$y_{t+1} = Ay_t + \eta_{t+1} + n$$

where

$$\eta_{it+1} = \ell n \lambda_{it+1}.$$

We have seen that the intertemporal optimization problem of this example led to the difference equation (8) which generates the sequence of y's. When we introduce randomness into the model, by means of stochastic yields, for example, then the difference equation becomes stochastic difference equation (9) which generates a sequence of random variables, i.e., a time series.

This difference equation is in the state space i.e., in a Markovian model form because it is a first order difference equation for the state vector. How does it relate to models more familiar to econometricians? Is it an AR, MR or ARMA model? We can answer this question easily by applying the Cayley-Hamilton theorem to eliminate the matrix A from the dynamic relations between y's and exogenous noises. (See Aoki [1976; p.45], for example.) This theorem states that the matrix A being 2 by 2, A^2 can be expressed as a linear combination of A and I, i.e., $A^2 = -\alpha A - \beta I$ for some constant α and β. The dynamic equation is $y_{t+1} = Ay_t + v_t$ where A is 2 by 2. Advance t by one to note that $y_{t+2} = Ay_{t+1} + v_{t+1} = A(Ay_t + v_t) + v_{t+1} = A^2y_t + Av_t + v_{t+1}$. Multiplying y_{t+1} and y_t by the constants α and β respectively, and add them to y_{t+2} to obtain

$$y_{t+2} + y_{t+1} + \beta y_t = (A^2 + \alpha A + \beta I)y_t + v_{t+1} + (A + \alpha I)v_t$$
$$= v_{t+1} + (A + \alpha I)v_t.$$

This is an ARMA model involving vector processes $\{y_t\}$ and $\{v_t\}$. The converse procedure is also possible, i.e., given any model in AR, MA, ARMA or ARIMA forms etc., they can be converted to state space or to Markovian model forms. As we show elsewhere the state space or Markovian model representaion and ARMA-like representation are equivalent.

Note that the elements of the matrix A are the parameters of the production function. The parameter θ_0, θ_1 and θ_2 characterize the utility function. The dynamics exhibit oscillatory behavior if eigenvalues of A are complex, or one peak may exist for a two-dimensional dynamics even when the two eigenvalues both have negative real parts.

Will this two sector model exhibit a hump-shaped multiplier profile said to be characteristic of the real output? The dynamic multiplier of (8) is given by $A^k n$. Using the spectral decomposition of A, we can write

$$A^k = \Sigma_i \lambda_i^k u_i v_i',$$

where λ_i is the eigenvalue corresponding to the right eigenvector u_i, and v_i' is its (row) left-hand eigenvector.

For example, the total output multiplier with an exogenous shock to the second sector is equal to $(1\ 1)A^k \begin{pmatrix} 0 \\ 1 \end{pmatrix} = \Sigma_{i=1}^2 \lambda_i^k (u_{i1} + u_{i2}) v_{i2}$, $k = 0, 1, \ldots.$ This is the multiplier time profile of exogenous shocks to the second sector. For the matrix A, the eigenvectors are

$$u_i = \begin{pmatrix} 1 \\ (\lambda_i - a_{11})/a_{12} \end{pmatrix}, \qquad i = 1, 2, \text{ and}$$

$$\begin{pmatrix} v_1' \\ v_2' \end{pmatrix} = \frac{a_{12}}{\lambda_2 - \lambda_1} \begin{pmatrix} (\lambda_2 - a_{11})/a_{12} & -1 \\ (a_{11} - \lambda_1)/a_{12} & 1 \end{pmatrix}.$$

Hence

$$u_{i1} + u_{i2} = (\lambda_i - a_{11} + a_{12})/a_{12},$$

$$v_1' \begin{pmatrix} 0 \\ 1 \end{pmatrix} = -a_{12}/(\lambda_2 - \lambda_1),$$

and

$$v_2' \begin{pmatrix} 0 \\ 1 \end{pmatrix} = a_{12}/(\lambda_2 - \lambda_1).$$

The multiplier profile is given by

$$m_t = (\lambda_2 - \lambda_1)^{-1} \{-(\lambda_1 - a_{11} + a_{12})\lambda_1^t + (\lambda_2 - a_{11} + a_{12})\lambda_2^t\}.$$

A sufficient condition for the series $\{m_t\}$ to exhibit a peak is to have $m_1 > 1$ because $m_0 = 1$ and $m_\infty = 0$, or $a_{22} + a_{12} > 1$. This condition may be interpreted as good 2 is productive as an input good. Although $a_{i1} + a_{i2} < 1$, $i = 1, 2$, the sum $a_{12} + a_{22}$ can very well be greater than one. For example $a_{11} = 0.3$, $a_{12} = 0.5$, $a_{21} = 0.2$ and $a_{22} = 0.6$ yields $a_{22} + a_{12} = 1.1 > 1$.

11.2 Quadratic Regulation Problems

Minimization of quadratic costs subject to linear dynamic constraints is often called LQ problems, and is basic in many intertemporal optimization formulation. This class of problems is basic partly because the LQ problems are analytically tractable and give us insight into structure of more general problems, while minimization of nonquadratic costs or inclusion of nonlinear constraints usually lead to analytically intractable problems. This fact alone justifies the study of the LQ problems. Furthermore, optimization problems with nonquadratic criteria and/or nonlinear dynamic constraints can often be iteratively approximated by a sequence of problems with quadratic costs and linear dynamic constraints. See Aoki [1962] for example. This is another reason for studying this class of intertemporal optimization problems.

This section discusses the LQ problems for continuous dynamic systems and discrete-time dynamics. See Canon et al. [1970] or Appendix A.16 for general statements of the first order necessary conditions for optimality for discrete time problems (discrete maximum principle), for example. Whittle [1982] has a readable treatment of the LQ problems for discrete time dynamics. The maximum principle for continuous time systems is discussed in a number of books, such as Lee and Markus [1967], Fleming and Rishel [1975] and Kamien and Schwarz [1981].

Discrete-time Systems

Dynamic Programming is a powerful conceptual tool for dealing with sequential decision problems, i.e., intertemporal optimization problems. Bellman's principle of optimality produces functional equations for optimal value functions of dynamic optimization problems. Unfortunately the functional equations must be solved numerically except for a few special cases. Linear dynamic systems with quadratic performance indices admits an explicit solution to the functional equations of Dynamic Programing.

Measuring state vectors and instrument vectors from appropriate references or base time paths, "regulation" or "tracking" problems are often formulated as follows:

Minimize $z_T'P_Tz + \Sigma_{\tau=t}^{T-1}W_\tau$

where $W_\tau = z'_\tau\theta z_\tau + x_\tau'R_\tau x_\tau$,

subject to the constraint

$$z_{\tau+1} = A_\tau z_\tau + B_\tau x_\tau.$$

Denote the optimal value of the criterion by $J_{t,T}(z_t)$. Bellman's principle of optimality yields the functional equation

(1) $$J_{t,T}(z_t) = \underset{x_t}{\text{Min}}\ \{W_t + J_{t+1,T}(z_{t+1})\}$$

where the minimization is with respect to the current choice vector x_t.

Note the terminal condition

$$J_{T,T}(z_T) = z_T'Pz_T.$$

This functional equation admits a solution of the form

(2) $$J_{t,T}(z) = z'\Pi_{t,T}z_T.$$

Clearly, Π_{TT} equals P.

To eliminate clutter of subscripts, we use a useful convention used by

Whittle [1982] of understanding by $(\ )_t$ that all subscripted variables in the brackets have the same subscript t, unless otherwise noted. For example, $(A_t + B_t C_{t+1})$ will be denoted by $(A + BC_{t+1})_t$.

Substituting (2) into (1), the functional equation now becomes

$$z_t' \Pi_{t,T} z_t = \operatorname*{Min}_x [z'Qz + x'Rx + (Az + Bx)' \Pi_{t+1,T} (Az + Bx)]_t .$$

The expression in the brackets may be written as a quadratic form $(z', x')\Pi \begin{pmatrix} z \\ x \end{pmatrix}$ where

$$\Pi = \begin{pmatrix} \Pi_{zz} & \Pi_{zx} \\ \Pi_{xz} & \Pi_{xx} \end{pmatrix} = \begin{pmatrix} Q + A'\Pi_{t+1,T}A & A'\Pi_{t+1,T}B \\ B'\Pi_{t+1,T}A & R + B'\Pi_{t+1,T}B \end{pmatrix}.$$

The matrix Π is symetric and non-negative definite and $\Pi_{xx} = R + B'\Pi_{t+1,T}B$ is positive definite. The minimizing value of x is determined by

$$\Pi_{xz} z_t + \Pi_{xx} x_t = 0$$

or

$$x_t = K_t z_t$$

where

$$K_t = -\Pi_{xx}^{-1} \Pi_{xz} .$$

The minimal value becomes

$$\Pi_{t,T} = \Pi_{zz} - \Pi_{zx} \Pi_{xx}^{-1} \Pi_{xz} .$$

Relabel $\Pi_{t,T}$ as V_t. Then this is a recursion for V_t. Because the corresponding equation for V_t is a differential equation known as a Riccati equation, the recursion for V_t is also called the Riccati equation of discrete-time LQ problems.

Restricting the class of decisions to be linear in z the recursion can also be written as

$$V_t = \operatorname*{Min}_K [Q + K'RK + (A + BK)' V_{t+1} (A + BK)] .$$

Bellman called it quasi-linear (because it is linear in V). He developed a method of approximation called quasi-linearization based on this equation. Also see Aoki [1968] who applied the quasi-linearization to obtain approximate solution to the Riccati equation.

As T goes to infinity, V_t or $\Pi_{t,T}$ approaches a constant if certain conditions are met. The limit satisfies the algebraic Riccati equation

$$V + Q + A'VA - A'VB(R + B'VB)^{-1}B'VA.$$

The optimal decision rule becomes

$$x_t = Kz_t$$

where

$$K = -(R + B'VB)^{-1}B'VA.$$

The same approach can handle problems in which a cross product term such as $zx_t'Sz_t$ is present in the criterion function.

We treat this problem by using another, and a short-cut method next. Consider a minimization problem with dynamic constraint

(3) $$z_{t+1} = Az_t + Bx_t$$
$$y_t = Cx_t$$

and the criterion function

$$J_{t,N} = \Sigma_t^{N-1}(y'Qy + x'Rx)_t$$

where

$$Q' = Q \geq 0, \ R' = R > 0.$$

Generate $\{V_t\}$ by

$$z_t'V_tz_t = z_{t+1}'V_{t+1}z_{t+1} + y_t'Qy_t + x_t'Rx_t.$$

Substitute (3) for z_{t+1} to rewrite the above as

(4) $$z_t'V_tz_t = [z'C'QCz + x'Rx + (Az + Bx)'V_{t+1}(Az + Bx)]_t$$

or

$$0 = z_t'(A'V_{t+1}A + C'QC - V_t)z_t + x_t'(R + B'V_{t+1}B)x_t$$
$$+ z_t'AV_{t+1}Bx_t + x_t'B'V_{t+1}AZ_t$$

$$= [(x - Kz)'(R + B'V_{t+1}B)(x - Kz)$$

$$+ z'(A'V_{t+1}A + C'QC - V - K'(R + B'V_{t+1}B)K)z]_t$$

where

$$K_t = -(R + B'V_{t+1}B)^{-1}B'V_{t+1}A.$$

In other words,

$$[y'Qy + x'Rx]_t = [(x - KZ)'(R + B'V_{t+1}B)(x - kz)$$

$$+ z'(A'V_{t+1}A + C'QC - V - K'B'V_{t+1}BK)z]_t + z_t'V_t z_t - z_{t+1}'V_{t+1}z_{t+1}.$$

Now related V_{t+1} to V_t by

(5) $$V_t = A'V_{t+1}A + C'QC - K_t'B'V_{t+1}BK_t$$

$$= A'V_{t+1}A + C'QC - A'V_{t+1}B(R + B'V_{t+1}B)^{-1}B'V_{t+1}A.$$

Then the criterion function is expressible in terms of V's by

$$\Sigma_{\tau=t}^{N}(y'Qy + x'Rx)_\tau = z_t'V_t z_t - z_N'V_N z_N + \Sigma_{T=t}^{N-1}(x_t - K_t z_t)'(R + B'V_{t+1}B)(x_t - K_t z_t).$$

Here $J_{t,N}$ is minimized by $x_t = K_t z_t$ and Min $J_{t,N} = z_t'V_t z_t$ by letting $V_N = 0$ as the terminal condition of the equation (5). Equation (5) is known as the (discrete) Riccati equation. We note that if $z_N'Tz_N + J_{t,N}$ is the cost function, then change of the terminal condition to $V_N = T$ is the only modification necessary.

The solution of a discrete-time regulator probelm with a slightly more general cost structure

(6) $$\text{Minimize } \Sigma_{t=0}^{\infty}(z_t', x_t')\begin{pmatrix} Q & S' \\ S & R \end{pmatrix}\begin{pmatrix} z_t \\ x_t \end{pmatrix}$$

with respect to $\{x_t\text{'s}\}$, subject to

(7) $$z_{t+1} = Az_t + Bx_t,$$

can be stated in terms of the Riccati equation

(8) $$P + A'PA + Q - (S + B'PA)'(R + B'PB)^{-1}(S + B'PA)$$

where $R + B'PB > 0$ is assumed.

As pointed out by Molinari [1975], (7) can be transformed by incorporating a reaction function

$$x_t = -Kz_t + v_t$$

into

(7') $$z_{t+1} = A_K z_t + Bv_t$$

where

$$A_K = A - BK.$$

Because the optimal solution is unique on the assumption that the controllability and observability conditions are met, the same P, which is the positive definite solution of (7), satisfies

$$P = A_K'PA_K + Q_K - (S_K + B'PA_K)'(R + B'PB)^{-1}(S_K + B'PA_K)$$

where

$$Q_K = Q - S'K - K'S + K'RK$$

$$S_K = S - RK.$$

The matrices Q_K and S_K are defined to keep the same cost expression.

11.3 Parametric Analysis of Optimal Solutions

Two or more distinct types of costs are often combined into a total cost function by assigning weights to each component of costs to produce a scalar-valued criterion functions for static optimization problems. Similarly, errors from different causes are jointed together with weights (such as inverses of error covariance matrices) to yield a scalar-valued criterion function in estimation problems. In such circumstances we want to know how sensitive optimal solutions are with respect to the weights in the criterion functions. Optimal estimation solutions often approach the least squares solutions as weights are taken to some limiting values. As an example, consider extracting an optimal secular growth time paths, $\{g_t\}$, from a given data, $\{y_t\}$, by minimizing the expression

$$\Sigma_t[(y_t - g_t)^2 + \lambda\{(g_t - g_{t-1}) - (g_{t-1} - g_{t-2})\}^2],$$

where two heterogeneous entities, i.e., the residuals $y_t - g_t$ and the second difference of the growth terms are combined together with weight λ to form an expression to be minimized. As λ approaches infinity, the optimal growth time path approaches the least squares fit of the data, $\{y_t\}$, by a linear trend term because g_t will approach $g_0 + \beta t$ for some constatnt β, Hodrick and Prescott [1981].

Such a parametric study is important in dynamic optimization problems as well. We wish to learn how the optimal solution behave as a function of the parameter in a criterion function, e.g., how large is the derivative (i.e., elasticity) of the optimal solution with respect to the parameter? Discrete-time problems turn out to be more cumbersome than continuous-time problems in answering this question. So we discuss the latter first.

Choice of Weighting Matrices

The spectral decomposition of dynamic matrix clearly show that the speeds of responses are determined by the eigenvalues while the shapes of the transient responses i.e., their time profiles are influenced by the eigenvectors. During the 1960's system theory has recognized, and used to advantage, the fact that feedback of state variables can be used to alter dynamics if the matrices A and B satisfy a certain rank condition (known as the stabilizability condition. See Wonham [1967] or Aoki [1976] for discussion of this condition). A feedback control rule or reaction function $x = Fz$ modifies the dynamic matrix A into A + BF. For stabilizable systems, the eigenvalues of the feedback dynamic matrix A + BF can be assigned arbitrarily subject only to the complex conjugacy condition. The eigenvalues of the closed loop systems (as feedback systems are often called) determine their speed of responses. The eigenvectors determine time profile or shapes of the transient responses. For systems with a single instrument, the associated engenvectors are also uniquely determined once the eigenvalues are assigned. Hence the speed of responses and the shapes of the transient responses are simultaneously determined for dynamic systems with single instruments.

Parametric dependence of eigenvalues can be examined by the method of root-locus for dynamic systems with single instrument and single target variable*. See Aoki [1976; Appendix B]. The analysis using the root-locus clearly show that if the transfer function is $n(s)/d(s)$ where $\deg d = p$ and $\deg n = q$, $q < p$, then $p - q$ of the eigenvalues go to infinity while p of the eigenvalues approach the roots of $n(s) = 0$ as certain system parameter approaches infinity.

Kwakernaak and Sivan [1972], Moore [1976], Harvey and Stein [1978] and others have generalized this result to multivariable cases and have also shown how the matrices Q and R in the quadratic criterion functions affect asymptotic properties of the optimized dynamics. In dynamic system with several instruments, specification of the feedback system eigenvalues i.e, the speeds of responses of $A + BF$ still leaves some freedom in choosing the associated eigenvectors or the shapes of the transient responses. This fact was established only in the middle of 1970's with the appearance of Moore [1976]. The next simple examples illustrate. Suppose $A = \begin{pmatrix} 0 & 1 \\ -\alpha & -\beta \end{pmatrix}$ and $B = \begin{pmatrix} 0 \\ 1 \end{pmatrix}$. Then with $F = (-f_1, -f_2)$, the feedback system dynamic matrix is $A + BF = \begin{pmatrix} 0 & 1 \\ -\gamma_1 & -\gamma_2 \end{pmatrix}$ where $\gamma_1 = \alpha + f_1$, $\gamma_2 = \beta + f_2$. The eigenvalues λ_1 and λ_2 are uniquely determined once γ_1 and γ_2 are given. The eigenvectors are $\begin{pmatrix} 1 \\ \gamma_1 \end{pmatrix}$, $\begin{pmatrix} 1 \\ \gamma_2 \end{pmatrix}$. Since γ_1 and γ_2 are uniquely

---

* Appendix A.8 shows how simple is the sensitivity analysis of dynamics when y and the decision variable are both scalar.

determined by f_1 and f_2 for a given A, the feedback matrix F uniquely determines the eigenvectors as well as eigenvalues. Next let the system has two instruments with $B = I_2$, and consider a reaction function with the feedback matrix F = $-\begin{pmatrix} f_1 & f_2 \\ f_3 & f_4 \end{pmatrix}$. The characteristic polynominal is $\lambda^2 + \gamma_1\lambda + \gamma_2$ as before where $\gamma_1 = \beta + f_1 + f_4$, $\gamma_2 = (\alpha + f_3)(1 - f_2) + f_1(\beta + f_4)$. The eigenvalues are $\begin{pmatrix} 1 \\ \lambda_1 + f_1 \end{pmatrix}$ and $\begin{pmatrix} 1 \\ \lambda_2 + f_1 \end{pmatrix}$ where λ_1 and λ_2 are the eigenvalues. They are still uniquely determined once γ_1 and γ_2 are given. Now, however, there are many ways for changing eigenvectors while keeping λ_1 and λ_2 constant because there are more than one way of specifying γ_1 and γ_2 by changing the elements of F. For example by varying f's to keep $f_1 + f_4$ and $f_3(1 - f_2) + f_1f_4$ constants, the eigenvalues remain the same while the eigenvectors change.

This lack of uniqueness or freedom to choose eigenvalue-eigenvector pairs for multiple-output systems with the closed-loop dynamic matrix A + BF where B is n by m and F is m by n means that transient behavior of the closed-loop system can be influednced by our choice of eigenvectors which, in turn, gives us a clue for choosing correct weights in the criterion function to produce desired transient behavior. This relation is best understood by examing dynamic behavior of closed-loop systems when cost associated with changing instruments approaches zero, i.e., control is getting cheaper. In discussing this problem we also comment on the instrument instability question. For fuller discussion of the instrument instability see Aoki [1976; Section 5.2].

Asymptotic behavior is analogous to that of a single-input single-output system. Aoki [1976; Appendix D] has emphasized the usefulness of the method of root-locus to study parametric dependence of the closed-loop eigenvalues as a parameter varies. There, some eigenvalues of feedback systems are shown to approach zeros of the transfer functions, while the remainder goes off to infinity following a well established asymptotes. Harvey and Stein [1978] established analogous results when targets and instruments are both vectors of the same dimension. We follow them in broad outline and examine the asymptotic behavior of multiple-input and multiple-output feedback systems.

The problem is to minimize a criterion

$$\int_0^\infty (y_t'Qy_t + \rho x_t'Rx_t)dt$$

for a system

$$\dot{z} = Az + Bx$$

$$y = Cz$$

where B is $(n \times r)$ and C is $(m \times n)$, $x = Fz$ is the best reaction function and examine the resultant system

$$\dot{z} = (A + BF)z$$

as $\rho \downarrow 0$.

The reaction function $x = Fz$ converts the dynamic system $\dot{z} = Az + Bx$ into the closed-loop or feedback system with dynamics $\dot{z} = (A + BF)z$. Assign a set of n eigenvalues of the dynamic matrix $A + BF$, subject to the complex conjugacy condition and all eigenvalues having negative real parts to make the matrix asymptotically stable. Moore [1976] showed that such a F exists if and only if (i) there exists a corresponding set of linearly independent eigenvectors v_i such that $(A + BF)v_i = \lambda_i v_i$, subject to the complex conjugacy condition, i.e., if $\lambda_j = \bar{\lambda}_i$, the $v_j = \bar{v}_i$, and (ii) the vector v_i is in the range space of N_{λ_i} where $\begin{pmatrix} N_\lambda \\ M_\lambda \end{pmatrix}$ spans the null space of $[\lambda I - A, B]$.* Such a matrix F is unique if rank B = m = dim y, i.e, if the dimensions of y and x

---

* There is a vector k_i such that $v_i = N_{\lambda_i} k_i$ and $-Fv_i = M_{\lambda_i} k_i$, $i = 1 \sim n$.

agree, because we can take B to be full rank without loss of generality.* Condition (ii) is the non-trivial condition of the two. The necessity of condition (ii) is easy to see. From $(A + BF)v_i = \lambda_i v_i$ follows $(\lambda_i I - A)v_i - BFv_i = 0$ or $\begin{pmatrix} v_i \\ -Fv_i \end{pmatrix} \in$ null space of $[\lambda_i I - A, B]$. This condition is conveniently expressed in terms of a matrix $T(\lambda)$ (called the return-difference matrix in the systems literature) as, $T(\lambda_i)\mu_i = 0$ where $\mu_i = FN_i$, and $T(\lambda) = I - F(\lambda I - A)^{-1}B$. (We also meet the return-difference matrices in identifying closed-loop systems in Chapter 12.) Since A + BF is uniquely determined by its (distinct) eigenvalues and eigenvectors (think of the spectral decomposition of A + BF), F is unique whenever the column vectors of B are linearly independent.

To establish sufficiency, suppose that a set of n linearly independent v_i, i = 1, ..., n, have been chosen subject to the complex conjugacy condition (i) and v_i is expressible as $v_i = N_{\lambda_i} k_i$. Hence $(\lambda_i I - A)N_i + BM_{\lambda_i} k_i = 0$. We next show that F is determined by the conditions $Fv_i = -M_{\lambda_i} k_i$, $i = i \sim n$. Granting this for the moment, we then establish $0 = (\lambda_i I - A)v_i - BFv_i$ or $(A + BF)v_i = \lambda_i v_i$, i.e., v_i is an eigenvector of A + BF with the eigenvalue λ_i. If all n eigenvalues are real, then v_i and $-M_{\lambda_i} k_i$ are all real. Hence the real matrix F can be solved out from

---

* It is known that unless $r \geqslant m$, a quadratic cost $\int_0^\infty (z'Qz + \rho x'Rx)dt$ can not be reduced to zero even if $\rho \downarrow 0$. In other words, if $r > m$, then the minimum of the cost as $\rho \downarrow 0$ has a positive limit $z_0'P_0z_0$ where $\rho_0 > 0$. This result obtained by Kwakernaak and Sivan [1972] can be transcribed for discrete time systems.

For this reason we examine the system where r = m with the additional assumption that B and C are full rank, i.e., rank (CB) = m. This can always be achieved. So it constitutes no real constraint on the problem we wish to examine here.

$$f[v_1, \ldots, v_n] = [w_1, \ldots, w_n], \qquad \text{where } w_i = -M_{\lambda_i} k_i,$$

as $F = [w_1, \ldots, w_n][v_1, \ldots, v_n]^{-1}$. When some λ's are complex conjugate, we need to manipulate the expressions to involve only real numbers by the usual transformation. We illustrate the procedure when $\lambda_2 = \bar{\lambda}_1$, $v_2 = \bar{v}_1$. Express $v_1 = \alpha + j\beta$. Then $v_2 = \alpha - j\beta$. Correspondingly $w_1 = \gamma + j\delta$ and $w_2 = \gamma - j\delta$. We have

$$[v_1, v_2, \ldots]\begin{pmatrix} \frac{1}{2} & -\frac{1}{2}j & 0 \\ \frac{1}{2} & \frac{1}{2}j & 0 \\ 0 & 0 & I \end{pmatrix} = [\alpha, \beta, v_3, \ldots],$$

$$[w_1, w_2, \ldots]\begin{pmatrix} \frac{1}{2} & -\frac{1}{2}j & 0 \\ \frac{1}{2} & \frac{1}{2}j & 0 \\ 0 & 0 & I \end{pmatrix} = [\alpha, \delta, w_3, \ldots],$$

and the equation to determine F now involves only real numbers $F[\alpha, \beta, v_3, \ldots] = [\alpha, \delta, w_3, \ldots]$.

The matrix T is related to the transfer functin by

(1) $\quad T'(-s)(\rho R)T(s) = \rho R + H'(-s)QH(s)$

where

$$H(s) = C(sI - A)^{-1}B, \text{ and}$$

when F is optimally chosen, i.e.,

$$F = -R^{-1}B'P/\rho$$

where P is the solution of the algebraic Riccati equation

(2) $\quad 0 = A'P + PA + C'QC - PBR^{-1}B'P/\rho.$*

From the vanishing of $T(\lambda_i)\mu_i$ follows that $H'(-\lambda_i)H(\lambda_i)\mu_i = 0$ or $H(\lambda_i)\mu_i = 0$ in the limit $\rho \downarrow 0$, becasue H(s) has no zero in the right half plane.

---

* Noting that $RF = -B'P$, expand the right hand side as $\rho R + B'(-sI - A')^{-1} \cdot pB + B'P(sI - A)^{-1}B + B'(-sI - A')^{-1}P(BR^{-1}B'/\rho)P(sI - A)^{-1}B = \rho R + B'(-sI - A')^{-1} \cdot [P(sI - A) + (-SI - A')P + P(BR^{-1}B'/\rho)P](sI - A)^{-1}B$ where the expression inside the square bracket equals $C'QC$ by (2). This establishes the equality of (1).

Because $(\lambda_i I - A)^{-1}B_{\mu_i}$ equals v_i, this condition is equivalently put as

$$Cv_i = 0, \qquad i = 1, \ldots, n-m,$$

since the null space of C is (n - m)-dimensional. The vectors v_i, $i = 1 \sim n-m$ span the null space of C. These (n - m) eigenvectors correspond to (n - m) eigenvalues that remain finite as $\rho \downarrow 0$. The remaining m eigenvalues go off to infinity as $\rho \downarrow 0$. To capture them let $\lambda_i = \lambda_i^{\infty}/\sqrt{\rho}$. Then expanding H(s) in a Laurent series and setting $s = \lambda_i$, (2) becomes

$$\begin{aligned} T'(-s)RT(s) &= \rho\{R - \frac{1}{\rho\lambda_i^2}(CB)'Q(CB) + \ldots\} \\ &= \rho\{R - \frac{1}{(\lambda_i^{\infty})^2}(CB)'Q(CB) + \ldots\}. \end{aligned}$$

Thus in the limit $\rho \downarrow 0$, the condition for the vecotr μ_i^{∞} becomes

$$R\mu_i^{\infty} = (\lambda_i^{\infty})^{-2}(CB)'Q(CB)\mu_i^{\infty}, \qquad i = 1, \ldots, m. \tag{3}$$

Define the $m \times m$ nonsingular matrix by

$$N_{\infty} = [\mu_1^{\infty}, \ldots, \mu_m^{\infty}].$$

Then (3) can be collectively written as

$$R^{-1}(CB)'Q(CB)N_{\infty} = N_{\infty}S_{\infty}^2 \tag{4}$$

where

$$S_{\infty} = \mathrm{diag}(\lambda_1^{\infty}, \ldots, \lambda_m^{\infty}).$$

This equation clearly establishes that the vectors μ_i^{∞}, $i = 1 \sim m$ are the eigenvectors of $R^{-1}(CB)'Q(CB)$. The matrix Q and R in the criterion function affects N_{∞} not as a simple ratio $R^{-1}Q$ but rather as $R^{-1}(CB)'Q(CB)$. Since only this ratio matters, we may take R as

$$R = (N_{\infty}S_{\infty}^2N_{\infty}')^{-1} \tag{5}$$

and let

$$(CB)'Q(CB) = (N_{\infty}N_{\infty}')^{-1} \tag{6}$$

because this choice preserves the key relation (4).

The consideration above suggests a coordinate system to examine the contribution to the quadratic cost. Let $T = [V^0, BN_\infty]$, and change variables to $z = Tw$, and $x = N_\infty v$.

By construction

$$AV^0 = V^0 S_0 - BN_0$$

where $S_0 = \mathrm{diag}(\lambda_1^0, \ldots, \lambda_{n-m}^0)$, where λ_i^0 is the finite eigenvalue as $\rho \downarrow 0$, and

$$N_0 = [\mu_1^0, \ldots, \mu_{n-m}^0].*$$

Let $T^{-1} = \begin{pmatrix} T_1 \\ T_2 \end{pmatrix}$. From $I_n = T^{-1}T$, we note that

$$T_1 V^0 = I_{n-m},\ T_1 BN_\infty = 0,$$

and

$$T_2 V^0 = 0 \text{ and } T_2 BN_\infty = I_m.$$

Noting that

$$T^{-1}AT = \begin{pmatrix} S_0 & A_{12}^0 \\ -N_\infty^{-1}N^0 & A_{22}^0 \end{pmatrix}, \qquad \text{for some } A_{i2}^0,\ i = 1, 2,$$

and

$$T^{-1}B = \begin{pmatrix} 0 \\ I_m \end{pmatrix},$$

the state equation for the new state vector w becomes

$$\begin{pmatrix} \dot{w}_1 \\ \dot{w}_2 \end{pmatrix} = \begin{pmatrix} S_0 & A_{12}^0 \\ -N_\infty^{-1}N^0 & A_{22}^0 \end{pmatrix} \begin{pmatrix} w_1 \\ w_2 \end{pmatrix} + \begin{pmatrix} 0 \\ I_m \end{pmatrix} v \tag{7}$$

and

$$y = CTw = CBN_\infty w_2.$$

The integrand of the cost function becomes, for our representation of R and Q in (5) and (6)

$$Y'Qy + \rho x'Rx = w_2'w_2 + v'S_\infty^{-2}v. \tag{8}$$

Equation (8) states that only the second subvector w_2, i.e., the subvector associated with fast modes (eigenvalues with large negative real

---

* When some of λ_i^0 are complex, the corresponding section of S_0 are $(\tau \times \tau)$ block diagonal submatrices.

values) contributes to the cost. Furthermore, it also reveals that the control cost, i.e, cost associated with changing instruments are weighted by the matrix S_∞^{-2}. The subvector w_1 represent relatively slowly decaying modes (compared with fast decaying modes of the subvector w_2) of the feedback system. Behavior of w_1 is determined by the zeros of the transfer function which are the diagonal elements of S_0.

We can bound the behavior of w_1 by

$$\|w_1(t)\| = \|e^{S_0 t} w_1(0) + \int_0^t e^{S_0(t-\tau)} A_{12}^0 w_2(\tau)d\tau\|$$

where the second term is bounded from above by

$$\|e^{S_0(t-\tau)} A_{12}^0 w_2(\tau)\| \leq \|A_{12}^0\|^2 w(0)'\Pi w(0)$$

where Π is the solution of

$$0 = H'\Pi + \Pi H + \begin{pmatrix} 0 & 0 \\ 0 & I \end{pmatrix} - \Pi\begin{pmatrix} 0 \\ I \end{pmatrix} S_\infty^2 (0,\ I)\Pi.$$

As $\rho \downarrow 0$ this matrix $\Pi \to 0$ if the numerator $\psi(s)$ of the transfer function $C(sI - A)^{-1}B$, i.e., $\psi(s) = |sI - A||C(sI - A)^{-1}B|$, has no zeros in Re $s > 0$, i.e., if the transfer function is of minimum phase. (Kwakernaak and Sivan [1972].)

The effects of w_1 on subvector w_2 eventually disappear and w_2 is essentially governed by

$$\dot{w}_2 = A_{21}^0 w_2(t) + v_t.$$

By examing the (2, 2) submatrix of the Riccati equation as $\rho \downarrow 0$, the dominant term of the matrix P is

$$P_{22} = I - \frac{1}{\rho} P_{22} S_\infty^2 P_{22}$$

or

$$P_{22} = \sqrt{e}\, S_\infty^{-1}.$$

The optimal feedback rule is

$$v = -(R^{-1}B'P/\rho)w \quad -(S_\infty^2/\rho)P_{22}w_2 = -(S_\infty/\sqrt{\rho})w_2$$

hence asymptotically w_2 is governed by

$$\dot{w}_2 = A^0_{22}w_2 - (S_\infty/\sqrt{\rho})w_2 = (A^0_{22} - S_\infty/\sqrt{\rho})w_2 = -(S_\infty/\sqrt{\rho})w_2$$

or

$$w_2(t) = e^{-(S_\infty/\sqrt{\rho})t}w_2(0)$$

and

$$z(t, \rho) = V^0 w_1 + BN_\infty w_2 \simeq BN_\infty e^{-(S_\infty/\sqrt{\rho})t}w_2(0).$$

With discrete time systems a similar analysis is possible, or the results may be translated via the bilinear transformation. See Section 3.3. The representation corresponding to (7) is less dramatic since m eigenvalues go only to the circle $|z| = 1$ rather than going off to ∞ as $\rho \downarrow 0$.

In discrete-version of this section, more of the eigenvalues go to infinity even when the cost of control approaches zero.

In a simpler situation of a single-input, single-output dynamics, $\chi_{t+1} = A\chi_t + bu_t$, $y_t = c\chi_t$ where $c = (c_0, c_1, \ldots, c_{n-\ell}, 0, \ldots, 0)$, and the criterion function $\Sigma_0^\infty(y_t'y_t + \rho u_t^2)$, the return difference matrix $T(z) = I + k'(zI - A)^{-1}b$ satisfies the factorization form

$$T'(z^{-1})(\rho + b'Pb)T(z) = \rho + b'(z^{-1}I - A')^{-1}cc'(zI - A)^{-1}b$$

from the optimal gain $u_t = -k'\chi_t$, $k' = \rho^{-1}b'P$ and P satisfies

$$P = APA' + c'c.$$

The eigenvalues of the closed-loop system is

$$|zI - A + bk'| = |zI - A|(1 + k'(zI - A)^{-1}b)$$

or

$$T(z) = |zI - A + bk'|/|zI - A|.$$

Here as ρ approaches zero, $(n - \ell)$ of the zeros of $T(z)$ approach the zeros of the tansfer function $c'(zI - A)^{-1}b$. The ℓ remaining zeros approach the origin. See Priel and Shaked [1983].

12 IDENTIFICATION

We want to select a model from a prescribed class of models which best "reproduces" observed data, given the same set of exogenous input sequences. This is the subject of identification.

Two notions of identifiability are found in the literature; consistency and uniqueness. Suppose that for a suitable parametrization of models, the parameter θ uniquely specifies a model within the class. We may then speak of the parameter as the model. Depending on the class of candidate models, the "ture" model or θ_0 may or may not be found in it. When it is, the convergence of the estimated parameter $\hat{\theta}$ to the true one is an issue. This is called consistency oriented identifiability (Wertz [1982]). Even when the true parameter value is not in the class, if each model in the class generates a distinct output sequences so that only one model or its parametric representation θ corresponds to a given input-output representation, then a "uniqueness oriented" identification can be examined. In other words, the examined issue is whether two models with different parameter values generate the same output sequences from the same input sequences, hence are indistinguishable or not.

Here we examine the latter notion, because statistical properties of estimating the parameters have been discussed extensively in the literature. Different parameter values must produce different probability distributions of data for the model to be called identifiable (Solo [1983]). Two models are observationally equivalent if the probability distributions of data are the same (in response to the same input sequences). If two observationally equivalent models are indeed the same then the model is identifiable. Taking the uniqueness view of identification, a function $g(\theta)$ of parameter vector θ is identifiable if equality of two probability distributions of data vector

y, $p(y|\theta_1) = p(y|\theta_2)$, implies that $g(\theta_1) = g(\theta_2)$, i.e., observationally equivalent models assign the same value to $g(\theta)$.

When only the first and the second moment information are used in identification, each model is then parametrized by mean of a vector and its covariance matrix. Two models in Markov or state space representation are indistinguishable if the covariance matrices are the same and the Markov parameters are identifiable, i.e., $H_i(\theta_1) = H_i(\theta_2)$ $i = 0, 1, \ldots$ because the impulse responses of state space models are completely specified by the set of Markov parameters $\{H_i\}$. In the ARMA representation $A_2(z) = M(z)A_1(z)$ and $B_2(z) = M(z)B_1(z)$ for some unimodular matrix $M(z)$ if and only if the Markov parameters coincide. Also recall that the Markov parameters are invariant with respect to similarity transformation in the state space, i.e., different choice of coordinate systems leave the Markov parameters invariant.

If a time series $\{y_t\}$ is mean-zero Gaussian, its covariance matrices $R_y(k) = E(y_{t+k}y_t')$ completely specifies the probability law for the y_t. Hence (θ_1, Q_1) and (θ_2, Q_2) are indistinguishable if and only if the covariance matrices are the same $R_y(k;\ \theta_1, Q_1) = R_y(k;\ \theta_2, Q_2)$. Even when the data are not Gaussian, they are indistinguishable if the above equality holds so long as we deal with its second-order properties.

A function $g(\theta)$ of the parameter θ is estimable if there exists a function of data y, $\phi(y)$, such that $g(\theta) = E_\theta(\phi(y))$. Estimable functions are identifiable because $P_{\theta_1}(y) = P_{\theta_2}(y)$ implies that $g(\theta_1) = g(\theta_2)$, if $g(\theta)$ is estimable. In the ARMA models, covariances $\{R_k(\theta)\}$ determine $P_\theta(y)$. Hence the coefficients in the AR polynominal is identifiable if and only if rank $\mathbb{H}(\theta) = p$ because $R_p(\theta_1) = R_p(\theta_2)$ implies $\mathbb{H}(\theta_1) = \mathbb{H}(\theta_2)$ and $r_p(\theta_1) = r_p(\theta_2)$.

12.1 Closed-Loop Systems

Time series are often related to each other. As an example consider two time seires $\{\eta_t\}$ and $\{\zeta_t\}$

(1)
$$\eta_t = P(L)\zeta_t + u_t,$$
and
$$\zeta_t = H(L)\eta_t + v_t,$$

where the matrices P(L) and H(L) are rational transfer matrices of L, and $\{u_t\}$ and $\{v_t\}$ are the exogenous disturbances. We say that the two time series are related by a closed-loop or feedback system because η_t is dynamically related to ζ_t which, in turn, is related to the original η_t completing or closing a loop.

The time series are directly related to the exogenous sequences by solving the simultaneous equation

$$\begin{pmatrix} I & -P(L) \\ -H(L) & I \end{pmatrix} \begin{pmatrix} \eta_t \\ \zeta_t \end{pmatrix} = \begin{pmatrix} u_t \\ v_t \end{pmatrix},$$

or

$$\begin{pmatrix} \eta_t \\ \zeta_t \end{pmatrix} = \begin{pmatrix} I & P(L) \\ H(L) & I \end{pmatrix} S \begin{pmatrix} u_t \\ v_t \end{pmatrix},$$

where

$$S = \mathrm{diag}(S_1(L), S_2(L))$$
$$S_1(L) = \{I - P(L)H(L)\}^{-1}$$

and

$$S_2(L) = \{I - H(L)P(L)\}^{-1}.$$

The matrix S is known as the return difference matrix in the systems literature and is known to be useful in stating various conditions for the closed-loop systems in terms of the original transfer matrices. (It also appears in stability and sensitivity analysis of closed-loop systems. See Section 11.3.)

Assume that the exogenous noises are related to stationary zero-mean white noise sequences with finite covariances

$$\begin{pmatrix} u_t \\ v_t \end{pmatrix} = \begin{pmatrix} N_1(L) & 0 \\ 0 & N_2(L) \end{pmatrix} \begin{pmatrix} a_t \\ b_t \end{pmatrix}, \quad \text{where } \operatorname{cov}\begin{pmatrix} a_t \\ b_t \end{pmatrix} = \Sigma = \operatorname{diag}(\Sigma_1, \Sigma_2).$$

The joint relations between η_t, ζ_t, u_t and v_t are stated then by the transfer function G(L):

$$\begin{pmatrix} \eta_t \\ \zeta_t \end{pmatrix} = G(L) \begin{pmatrix} a_t \\ b_t \end{pmatrix},$$

where

$$G(L) = \begin{pmatrix} G_{11}(L), & G_{12}(L) \\ G_{21}(L), & G_{22}(L) \end{pmatrix} = \begin{pmatrix} I & P(L) \\ H(L) & I \end{pmatrix} S \begin{pmatrix} N_1(L) & 0 \\ 0 & N_2(L) \end{pmatrix}.$$

The submatrices of G are related to the original transfer matrices by

$$G_{11}(L) = S_1(L)N_1(L), \quad G_{12}(L) = P(L)S_2(L)N_2(L),$$
(2)
$$G_{21}(L) = H(L)S_1(L)N_1(L), \quad G_{22}(L) = S_2(L)N_2(L).$$

By assumption the matrix G is rational and stable. Additionally we assume that there is no pole-zero cancellation so that the dimension of a minimal realization of a pair of matrices (P, N_1), call it n_1, and of (H, N_2), denoted by n_2, add up to $n = n_1 + n_2$, the dimension of a minimal realization of the matrix G.

The spectrum of the closed-loop system then exists, is rational and given by

$$S(z) = G(z)\Sigma G'(z^{-1}).$$

We say that the transfer functions P(L), H(L), $N_1(L)$ and $N_2(L)$ are recoverable if models can be constructed with transer functions $\tilde{P}$, $\tilde{H}$, $\tilde{N}_1$, and $\tilde{N}_2$ which satisfy the relations

$$H = \tilde{H},\ P = \tilde{P},\ \text{and}\ N_i(z)\ \Sigma N_i'(z^{-1}) = \tilde{N}_i(z)\ \Sigma\tilde{N}_i'(z^{-1}), \qquad i = 1, 2.$$

The transfer functions for an open-loop model can be solved out from the matrix G of (2):

$$\text{(3)} \qquad \begin{aligned} &P = G_{12}G_{22}^{-1}, \quad H = G_{21}G_{11}^{-1} \\ &N_1 = G_{11} - G_{12}G_{22}^{-1}G_{21}, \quad N_2 = G_{22} - G_{21}G_{11}^{-1}G_{12}. \end{aligned}$$

Suppose that (A, B, C) is a minimal realization of G, i.e.,

$$G(z) = I + C(zI - A)^{-1}B$$

where

$$C = \begin{pmatrix} C_1 \\ C_2 \end{pmatrix}, \text{ and } B = [B_1, B_2],$$

or

$$\begin{aligned} G_{11} &= I + C_1(zI - A)^{-1}B_1, \\ G_{12} &= C_1(zI - A)^{-1}B_2, \\ G_{21} &= C_2(zI - A)^{-1}B_1, \\ G_{22} &= I + C_2(zI - A)^{-1}B_2. \end{aligned}$$

Then, a Markov model for the two time series is given by

$$w_{t+1} = Aw_t + B\varepsilon_t,$$

$$\begin{pmatrix} \eta_t \\ \zeta_t \end{pmatrix} = Cw_t + \varepsilon_t,$$

where

$$\varepsilon_t = \begin{pmatrix} a_t \\ b_t \end{pmatrix}.$$

In (3), we note that

$$G_{11}^{-1} = I - C_1(zI - A + B_1C_1)^{-1}B_1,$$

and

$$G_{22}^{-1} = I - C_2(zI - A + B_2C_2)^{-1}B_2.$$

FHence the transfer function P is given by

$$P = G_{12}G_{22}^{-1} = G_1(zI - A)^{-1}B_2 - C_1(zI - A)^{-1}B_2C_2(zI - A + B_2C_2)^{-1}B_2$$

$$= C_1(zI - A)^{-1}[zI - A + B_2C_2 - B_2C_2](zI - A + B_2C_2)^{-1}B_2$$

$$= C_1(zI - A + B_2C_2)^{-1}B_2,$$

and the transfer function H becomes

$$H = G_{21}G_{11}^{-1} = C_2(zI - A + B_1C_1)^{-1}B_1.$$

Because the matrix $G(z)$ is minimum phase by assumption, $G(z)^{-1}$ has all poles in $|z| \leqslant 1$. Then, the eigenvalues of $A - BC$ lie inside $|\lambda| \leqslant 1$, because $G(z)^{-1} = I + C(zI - A + BC)^{-1}B$, and the poles of $G(z)^{-1}$ are the roots of $|zI - A + BC| = 0$.

Hence

$$N_1^{-1} = I - C_1(B_1C_1 + zI - A + B_2C_2)^{-1}B_1$$

has all poles inside $|z| \leqslant 1$.

Similarly N_2^{-1} also has all poles inside $|z| \leqslant 1$.

We state this as

<u>Fact</u> The transfer function N_1 and N_2 are of minimum phase if $G(z)$ is minimum phase.

12.2 Identifiability of a Closed-Loop System

Here we follow Solo [1983] in establishing the identifiability of the autoregressive part of the closed-loop sytem.

Write (1) as

$$\eta_t - P\zeta_t = u_t = N_1 a_t$$

or

$$a_t = N_1^{-1}(\eta_t - P\zeta_t)$$
$$= (N_1^{-1}, -N_1^{-1}P)\begin{pmatrix}\eta_t\\ \zeta_t\end{pmatrix}.$$

For $\{a_t\}$ to be stationary, N_1^{-1} and $N_1^{-1}P$ are necessarily stationary, i.e., it is necessary that N_1 is minimum phase and $N_1^{-1}P$ be stable. Let θ_0 be the true parameter value of the autoregressive part of the closed-loop system. Then a fortiori, $N_1(\theta_0)$ is minimum phase and $N_1^{-1}(\theta_0)P(\theta_0)$ is stable.

Let

$$[P(\theta_0), N_1(\theta_0)] = A_{f_0}^{-1}(L)[B_{f_0}(L), C_{f_0}(L)].$$

Then $z^{p_f}C_{f_0}(z^{-1}) = 0$ has all roots < 1.* The criterion function $L(\theta) = E(e_t^2(\theta))$, where $e_t(\theta) = N_1(\theta)^{-1}(\eta_t - P(\theta)\zeta_t)$ is a-idendifiable if $L(\theta) = L(\theta_0)$ implies that $\theta = \theta_0$. Note that $e_t(\theta_0) = a_t$, and we next establish $L(\theta) \geqslant \sigma_1^2 = E(a_t^2)$, $\forall\theta$. To see this, substitute the system relation into

$$e_t(\theta) = N_1^{-1}[I, -P]\begin{pmatrix}\eta_t\\ \zeta_t\end{pmatrix}$$
$$= N_1^{-1}[I, -P]\begin{pmatrix}I & -P\\ -H & I\end{pmatrix}_0^{-1}\begin{pmatrix}N_1 & 0\\ 0 & N_2\end{pmatrix}\begin{pmatrix}a_t\\ b_t\end{pmatrix}$$
$$= N_1^{-1}[I, -P]\begin{pmatrix}(S_1N_1)_0 & (PS_2N_2)_0\\ (HS_1N_1)_0, & (S_2N_2)_0\end{pmatrix}\begin{pmatrix}a_t\\ b_t\end{pmatrix}$$
$$= a_t + T_1a_t + T_2b_t$$

where

$$T_1 = N_1^{-1}N_1^0 - I + N_1^{-1}(P^0 - P)(HS_1N_1)_0$$

and

$$T_2 = N_1^{-1}(P^0 - P)(S_2N_2)_0.$$

---

* If $a(L)\ a_0 + a_1L + \ldots + a_pL^p$, then $z^pa(z^{-1}) = a_0z^p + a_1z^{p-1} + \ldots + a_p$.

Note that $T_1(\theta_0) = 0$ and $T_2 = (\theta_0) = 0$. If $S_1^0(0) = 1$ and $S_2^0(0) = 1$, then T is strictly causal hence $L(\theta) \geqslant E(a_t^2)$. Thus $S^0(0) = I$ is necessary for the identifiability of the closed-loop system.

13 TIME SERIES FROM RATIONAL EXPECTATIONS MODELS

Expected future values of relevant endogenous and exogenous variables must be incorporated in rational economic decisions. Time series are governed, then, by a class of difference equations which involve conditionally expected values of future y's as well as current and past y's: a class we have not discussed so far. We follow Gourieroux et al. [1979] to characterize completely the solutions of a first order difference equation for y in which y_t and a one-step ahead prediction term $y_{t+1|t}$ appear

(1) $$y_t = ay_{t+1|t} + u_t$$

where a is a known scalar and $\{u_t\}$ is a mean-zero weakly stationary stochastic process. The symbol $y_{t+1|t}$ denotes the conditional expectation of y_{t+1} given an information set I_t where $I_t = \{\varepsilon_t, \varepsilon_{t-1}, \ldots\}$. Equations of the form (1) arise in many economic models. See Aoki and Canzoneri [1979] for the solution method in which terms related to $y_{t|t-1}$ rather than $y_{t+1|t}$ appear. As an example leading to dynamics of the type (1), suppose that the money demand function (in a high inflation economy) is specified by $m_t^d - p_t = \alpha(p_{t+1|t} - p_t)$ and the money supply is $m_t^s = \mu_t$ where p_t is the logarithm of price level. Then $p_t = ap_{t+1|t} + u_t$ where $a = \alpha/(\alpha - 1)$ and $u_t = \mu_t/(1 - \alpha)$. Here $p_{t+1|t} - p_t$ is a proxy for the interest rate because the expected inflation rate completely dominates any other effects in a high inflation economy.

We consider three possibilites: (i) when u_t is related to a basic stochastic process $\{\varepsilon_t\}$ by a MA process, (ii) by an AR process, and (iii) by an ARMA process. First, to obtain the solution of (1), we need a particular solution of the inhomogeneous part and general solutions of the homogeneous part: $y_t = ay_{t+1|t}$. The general solutions of (1) is related to a martingale.

This can be seen by converting $y_t = ay_{t+1|t}$ into $a^t y_t = a^{t+1} y_{t+1|t}$, and defining Z_t to be $a^t y_t$. Then this equation is the same as the definition of a martingale $E(Z_{t+1}|I_t) = Z_t$. Hence a general solution is of the form $y_t = a^{-t}Z_t$ for any martingale Z_t. Denote a particular solution of (1) by y_t^p, and let y_t^h be a general solution of (1). Adding these two together, $y_t^p + y_t^h$ satisfies (1).

This "superposition" principle also works with respect to the u_t specification. Suppose $u_t = \xi_t + \eta_t$ where ξ's and η's are mutually independent mean-zero stochastic processes. Then a particular solution for (1) can be made up as a sum of two separate particular solutions of (1), one with $u_t = \xi_t$, and the other with $u_t = \eta_t$ as disturbances. This is because $y_t^\xi = aE(y_{t+1}^\xi|\xi^t) + \xi_t$ and $y_t^\eta = aE(y_{t+1}^\eta|\eta^t) + \eta_t$ can be added together, because $E(y_{t+1}^\xi|\xi^t) = E(y_{t+1}^\xi|\xi^t,\eta^t)$ and $E(y_{t+1}^\eta|\eta^t) = E(y_{t+1}|\eta^t, \xi^t)$ by the independence of ξ^t and η^t, where $\xi^t = \{\xi_t, \xi_{t-1}, \ldots\}$ and similarly for η^t. Hence $y_t = y_t^\xi + y_t^\eta$.

A method of undetermined coefficients provides a basic procedure for solving (1) if the exogenous noises are independent. First, we illustrate it step-by-step. After a few practice examples, we can bypass many intermediate steps and proceed more directly to the solutions.

13.1 Moving Average Processes

Suppose now that u_t is MA(q), $u_t = C(L)\varepsilon_t = \varepsilon_t + C_1\varepsilon_{t-1} + \ldots + C_q\varepsilon_{t-q}$ where ε_t is a mean zero white noise process with unit variance. We assume that all the roots of $C(L) = 0$ lie outside the unit circle. Because of linearity of (1) and independence of ε's, we look for a particular solution to the equation

(2) $$y_t^i = ay_{t+1|t}^i + \varepsilon_{t-i}, \qquad i = 0, 1, \ldots, q.$$

Then, a particular solution $\Sigma_{i=1}^{q} C_i y_t^i$ satisfies (1). Here the conditioning variables are $\varepsilon^t = (\varepsilon_t, \varepsilon_{t-1}, \ldots)$ which are common to all i.

Hypothesize a solution to (2) to be given by

$$y_t^i = \alpha_0 \varepsilon_t + \alpha_1 \varepsilon_{t-1} + \ldots + \alpha_i \varepsilon_{t-i}$$

where α's are to be determined by substituting this hypothesized solution form into (2). Then advancing t by one in the above equation and projecting the resulting expression on the subspace spanned by ε^t we obtain

$$\alpha_0 \varepsilon_t + \alpha_1 \varepsilon_{t-1} + \ldots + \alpha_i \varepsilon_{t-i} = a(\alpha_1 \varepsilon_t + \ldots + \alpha_i \varepsilon_{t+1-i}) + \varepsilon_{t-i}.$$

Comparing the coefficient of ε_{t-j} with that on the right hand side, j = i, i-1, ..., 0, we determine that

$$\alpha_i = 1$$

$$\alpha_{i-1} = a$$

$$\vdots$$

$$\alpha_0 = a^i$$

i.e.,

$$y_t^i = T_i(L)\varepsilon_t$$

where

$$T_i(L) = a^i + a^{i-1}L + \ldots + L^i.$$

Consequently, a particular solution of (1) is

$$y_t = T(L)\varepsilon_t$$

where

$$T(L) = \sum_{i=0}^{q} C_i T_i(L).$$

To express y_t in terms of u_t, multiply both sides by C(L)

$$C(L)y_t = T(L)C(L)\varepsilon_t = T(L)u_t.$$

By assumption, the zeros of C(L) all lie outside the unit circle so 1/C(L) is a well-defined causal filter. We obtain a particular solution

$$y_t = \{T(L)/C(L)\}u_t.$$

This derivation does not reveal how T(L) relates to C(L), if at all. An alternative procedure which we later discuss tells us that

$$T(L) = C(a) + L\{C(L) - C(a)\}/(L - a).$$

This can be verified by substitution.

We now switch from L to z-variable, $L = z^{-1}$. The MA polynomial is

$$C(z^{-1}) = 1 + c_1 z^{-1} + \ldots + c_q z^{-q} = z^{-q}(z^q + c_1 z^{q-1} + \ldots + c_q).$$

So, in terms of the z-variable, all the finite zero of $C(z^{-1})$ lie inside the unit circle.

Now, hypothesize a particular solution of the form

$$y_t = \{\alpha + z^{-1}\gamma(z^{-1})\}\varepsilon_t$$

and see if α and $\gamma(z^{-1})$ exist that satisfy (1). Advance t by one in the above and take its conditional expectation $y_{t+1|t} = \gamma(z^{-1})\varepsilon_t$. Substitute this into (1) to obtain a relation

$$\{\alpha + (z^{-1} - a)\gamma(z^{-1})\}\varepsilon_t = C(z^{-1})\varepsilon_t.$$

Setting z^{-1} to a, we see that

$$\alpha = C(a).$$

Then $\gamma(\cdot)$ must be given by

$$\gamma(z^{-1}) = \{C(z^{-1}) - C(a)\}/(z^{-1} - a).$$

The right hand side is analytic in z^{-1}. For $\gamma(z^{-1})\varepsilon_t$ to be a well-defined back-shift-invariant subspace in the Hilbert space of random variables, $\gamma(z^{-1})$ must be analytic in $|z| \leqslant 1$ and have zeros inside the unit circle.

13.2 Autoregressive Processes

Here $I_t = \{u_t, u_{t-1}, \ldots\} = \{\varepsilon_t, \varepsilon_{t-1}, \ldots\}$. Let $\phi(L)u_t = \varepsilon_t$ where $\phi(L) = 1 + a_1 L + \ldots + a_p L^p$ with all zeros outside the unit circle. The polynomial $a(z^{-1})$ then has all finite zeros inside the unit circle. Try a solution of the form

(3) $$y_t = b(L)u_t$$

where

$$b(L) = b_0 + b_1 L + \dots + b_{p-1}L^{p-1}.$$

The conditional expectation then becomes

$$y_{t+1|t} = b_0 u_{t+1|t} + \beta(L)u_t$$

where

$$\beta(L) = b_1 + b_2 L + \dots + b_{p-1}L^{p-2}.$$

The conditional expectation $u_{t+1|t}$ is calculated analogously as $u_{t+1|t} = -\alpha(L)u_t$, because $u_{t+1} + \alpha(L)u_t = \varepsilon_{t+1}$ where $\alpha(L) = a_1 + a_2 L + \dots + a_p L^{p-1}$. Hence

(4) $$y_{t+1|t} = \{\beta(L) - b_0\alpha(L)\}u_t.$$

Substituting (3) and (4) into (1), we observe

$$b(L)u_t = a\{\beta(L) - b_0\alpha(L)\}u_t + u_t$$

If the polynominal $b(\cdot)$ is chosen to satisfy

$$b_0 + \beta(L)(L - a) + ab_0\alpha(L)^{-1} = 0$$

identically in L, then (3) is a particular solution.

Setting L to a in the above, the constant b_0 is equal to

$$b_0 = 1/\{1 + a\alpha(a)\} = 1/\phi(a)$$

if $\phi(a)$ is not zero. Assuming this for now, the polynomial $\beta(L)$ is determined then

$$\beta(L) = \{1 - \phi(L)/\phi(a)\}/(L - a),$$

Hence

$$b(L) = b_0 + L\ (L)$$
$$= \{1/\phi(a)\}\{1 - L\,\frac{\phi(L) - \phi(a)}{L - a}\ .$$

We can rewrite (3) as

$$\phi(L)y_t = b(L)\varepsilon_t$$

hence y_t is an ARMA(p, p-1).

When $\phi(a)$ is zero, the trial solution

$$y_t = \{L\beta(L) + t(\gamma_0 + L\Gamma_0/L)\}u_t$$

where $\deg \Gamma_0 = p - 2$ works.

If $p = 1$, then $b_0 = 0$ and $\Gamma_0(\cdot)$ is zero. If 1 is a root of $\phi(\ )$ of multiplicity d then y_t is an ARIMA (p-d, d, p-1). See Monfort et al. [1982].

13.3 ARMA Models

Consider

$$y_t = ay_{t+1|t} + u_t$$

where

$$\phi(L)u_t = C(L)\varepsilon_t$$

where the root of ϕ and C all lie outside the unit circle.

Multiply the model by $\phi(L)$ to render it as

$$\begin{aligned}(5)\qquad \phi(L)y_t &= a\phi(L)y_{t+1|t} + \phi(L)u_t \\ &= a\phi(L)y_{t+1|t} + C(L)\varepsilon_t.\end{aligned}$$

Introduce an auxiliary variable η_t by

$$\eta_t = \phi(L)y_t.$$

Then (5) is a first order difference equation for

$$\eta_t = a\eta_{t+1|t} + C(L)\varepsilon_t$$

which is the MA form discussed above. Its particular solution has been derived as $\eta_t = T(L)\varepsilon_t$ where $T(L) = C(a) + L\{C(L) - C(a)\}/(L - a)$. Hence

$$\begin{aligned}(6)\qquad y_t &= \{T(L)/\phi(L)\}\varepsilon_t = \{T(L)/C(L)\}\{C(L)/\phi(L)\}\varepsilon_t \\ &= \{T(L)/C(L)\}u_t.\end{aligned}$$

Thus a form $C(L)y_t = T(L)u_t$ is suggested as a possible solution, where $\deg T = \max(q, p-1)$.

We need $u_{t+1|t}$ in calculating $y_{t+1|t}$. Write u_{t+1} as

$$\begin{aligned} u_{t+1} &= \{C(L)/\phi(L)\}\varepsilon_{t+1} \\ &= (1 + C(L)/\phi(L) - 1)\varepsilon_{t+1} \\ &= \varepsilon_{t+1} + \{C(L)/\phi(L) - 1\}\varepsilon_{t+1}. \end{aligned}$$

Hence

$$\begin{aligned} u_{t+1|t} &= \frac{1}{L}\left(\frac{C(L)}{\phi(L)} - 1\right)\varepsilon_t \\ &= \frac{1}{L}\left(\frac{C(L)}{\phi(L)} - 1\right)\cdot\frac{\phi(L)}{C(L)}u_t \\ &= \frac{1}{L}\left(1 - \frac{\phi(L)}{C(L)}\right)u_t. \end{aligned}$$

Then advancing t by one in (6), and adding and subtracting an undetermined constant τ_0, we express

$$\begin{aligned} y_{t+1} &= \{T(L)/C(L)\}u_{t+1} \\ &= \{\tau_0 + \frac{1}{L}\left(\frac{T(L)}{C(L)} - \tau_0\right)L\}u_{t+1} \\ &= \tau_0 u_{t+1} + \frac{1}{L}\left(\frac{T(L)}{C(L)} - \tau_0\right)u_t, \end{aligned}$$

hence

$$y_{t+1|t} = \tau_0 u_{t+1|t} + \frac{1}{L}\left(\frac{T(L)}{C(L)} - \tau_0\right)u_t.$$

Substituting this into the original equation

$$T(L)/C(L)\ u_t = \frac{a\tau_0}{L}\left(1 - \frac{\phi(L)}{C(L)}\right)u_t + \frac{a}{L}\left(\frac{T(L)}{C(L)} - \tau_0\right)u_t + u_t$$

or

(7) $$T(L)/C(L) = \frac{a\tau_0}{L}(1 - \phi(L)/C(L)) + \frac{a}{L}\left(\frac{T(L)}{C(L)} - \tau_0\right) + 1.$$

Letting $L = a$, τ_0 must satisfy

$$T(a)/C(a) = \tau_0(1 - \phi(a)/C(a)) + (T/a)/C(a) - \tau_0) + 1$$

or if $\phi(a) \neq 0$, then $\tau_0 = C(a)/\phi(a)$.

Substituting this back into (7) determines

$$T(L) = \frac{1}{L - a}\{LC(L) - a\frac{C(a)}{\phi(a)}\phi(L)\}.$$

As pointed out by Monfort et al. this method is superior to the method of Blanchard [1979] which works only in subcases, as the next example shows. In the example Blanchard's method works only if $|a\varphi_1| < 1$.

13.4 Examples

Example 1 Suppose

$$u_t + \varphi_1 u_{t-1} = \varepsilon_t.$$

Then Blanchard's method solves this equation by successively advancing t

$$u_{t+1} = -\varphi_1 u_t + \varepsilon_{t+1}.$$

Hence $u_{t+1|t} = -\varphi_1 u_t$. Similarly $u_{t+2} = -\varphi_1 u_{t+1} + \varepsilon_{t+2}$. Hence $u_{t+2|t} = -\varphi_1 u_{t+1|t} = (-\varphi_1)^2 u_t$ and $u_{t+i|t} = (-\varphi_1)^i u_t$ in general $i > 0$. Hence

$$y_t = \sum_{i=0}^{\infty} (-a\varphi_1)^i u_t$$

converges if and only if $|a\varphi_1| < 1$. The Monfort procedure shows that $y_t = u_t/(1 + \varphi_1 a)$ as a particular solution always unless $1 + \varphi_1 a \neq 0$.

Example 2 Consider a simple model of a closed economy given by

(8) $$y_t = -\sigma\{i_t - (p_{t+1|t} - p_t)\} + \eta_t,$$

(9) $$y_t = a(p_t - p_{t|t-1}) + s_t,$$

(10) $$\mu_t = -ki_t + p_t + y_t.$$

Equation (8) is the aggregate demand equation. The aggregate supply function is given by (9). Behind (9) is a wage contracting story. The demand for real balances is given by (10), where the price index term drop out from both sides by assuming unit income elasticity of demand for real balances.

Solve (10) for i_t

$$i_t = \frac{-\mu_t + p_t + y_t}{k}.$$

Equating (8) with (9) we obtain the dynamic equation for the price time series:

$$\text{(11)} \qquad \pi_1 p_t - \pi_2 p_{t|t-1} - p_{t+1|t} = (1+\sigma/k)s_t - \eta_t - (\sigma/k)\mu_t$$

where

$$\pi_1 = a(1+\sigma/k) + \sigma(1+1/k), \quad \pi_2 = a(1+\sigma/k).$$

Suppose the noises on the right hand side is specified by $\Sigma_{i=1}^{3} f_j(L)\varepsilon_t^j$. Postulate that the solution is of the form $p_t = \Sigma_j(\gamma_j + Lh_j(L))\varepsilon_t^i$. Then $p_{t|t-1} = \Sigma_j Lh_j(L)\varepsilon_t^j$. Because $p_{t+1} = \Sigma_j(\gamma_j + Lh_j(L))\varepsilon_{t+1}^j$, we note that $p_{t+1|t} = \Sigma_j h_j(L)\varepsilon_t^j$. Substituting the postulated solution form into (11), we find that γ_j and $h_j(z)$ must satisfy (See Futia [1979, 1981]) the following relation

$$\pi_1\pi_j + (\pi_1 z - \pi_2 z - \sigma)h_j(z) = f_j(z).$$

Let λ be the root of $(\pi_1 - \pi_2)z - \sigma = 0$. Then we solve for γ_j by

$$\gamma_j = \frac{1}{\pi_1} f_j(\lambda)$$

when

$$\lambda = \sigma/(\pi_1 - \pi_2) = (1+1/k)^{-1},$$

and h_j is given by

$$h_j(z) = (f_j(z) - \pi_1\gamma_j)/(\pi_1 z - \pi_2 z - \sigma) = \{f_j(z) - f_j(\lambda)\}/[(\pi_1 - \pi_2)z - \sigma].$$

Note, however, that h_j will not be analytic inside the unit disc $|z| < 1$, unless

$$f_j(z) - f_j(\lambda) = (z - \lambda)^{\ell} g_j(z), \; g_j(\lambda) \neq 0, \quad \ell \geq 1$$

or

$$f_i(z) = c_j + (z - \lambda)^{\ell} g_j(z)$$

for some $c_j \neq 0$.

Some particular cases obtain by specializing $f_j(\cdot)$: Let $f_i(z) = c_j$. Then $\gamma_j = c_j/\pi_1$ and $h_j = 0$; If $f_j(z) = c_j + (z - \lambda)d_j$, then $\gamma_j = c_j/\pi_1$ and $h_j = d_j$.

Example 3 A simple N-sector stochastic employment model

The next example highlights the role of (disparate) information in generating serial correlations even when exogenous disturbances are serially uncorrelated. We do not wish to imply, however, that this is the only or the most important source for business cycles. Nevertheless this example is interesting because it does illustrate one often overlooked source of serial correlations.

Consider an economy with N sectors. The i-th sector employment level is related to the sector output by

(12) $$y_t^i = L_t^i,$$

and changes with time according to

(13) $$L_{t+1}^i = (1 - \delta)L_t^i + \theta E_t^i(y_{t+1}^i - y_{t+1}^a) + g_{t+1}^i$$

where the average of y's is defined by

$$y_{t+1}^a = \frac{1}{N}\Sigma_{j=1}^N y_{t+1}^j,$$

and g_t^i is an exogenous disturbance to be specified presently. The symbol E_t^i denotes the conditional expectation $E(\cdot|I_t^i)$ where I_t^i is the information set of sector i. Equation (13) describes processes of labor movements. Labor moves to a sector with higher than average prospect of employment which, according to (12) is equivalent to the sector with higher than (national) average output.* Substituting (1) into (2) and defining L_{t+1}^a as $\Sigma_{j=1}^N L_{t+1}^i/N$, we rewrite (13) as

(14) $$L_{t+1}^i = (1 - \delta)L_t^i + \sigma\theta E_t^i \ell_{t+1}^i + g_{t+1}^i$$

where we denote the deviation of the i-th sector employment from the average by ℓ_{t+1}^i, i.e., $\ell_{t+1}^i = L_{t+1}^i - L_{t+1}^a$.

Averaging this equation over N sectors, we obtain the dynamics for the aggregate or macro-system:

---

* Labor and output may be interpreted in per capital stock. Then variables measured from some trend (growth) path will be related by (12) and (13).

(15) $$L^a_{t+1} = (1-\delta)L^a_t + \frac{\sigma\theta}{N}\Sigma_j E^j_t \ell^j_{t+1} + g^a_{t+1}$$

and where g^a_t is the average disturbance defined to be $\Sigma_j g^j_t/N$.

Taking the difference of (14) and (15) we note that

(16) $$\ell^i_{t+1} = (1-\delta)\ell^i_t + \sigma\theta(E^i_t L^i_{t+1} - \frac{1}{N}\Sigma_j E^j_t \ell^j_{t+1}) + g^i_{t+1} - g^a_{t+1}.$$

The total system dynamics are described by the average behavior (15) and deviations from the average (16).

When we sum (16) over i we obtain $\Sigma\ell^i_{t+1} = (1-\delta)\Sigma\ell^i_t$. Hence if $\Sigma_j \ell^j_0 = 0$, then $\Sigma_j \ell^j_{t+1} = 0$ for all $t \geq 0$. Even if $\Sigma_j \ell^j_0 \neq 0$, the magnitude $|\Sigma\ell^i_t|$ monotonically converges to 0 as t goes to infinity. We assume that $\Sigma\ell^i_t = 0$ for all $t \geq 0$. We follow Futia [1979] and define a microequilibrium to be a collection of covariance stationary stochastic processes $\{\ell^i_t\}$ such that $E^i_t \ell^i_t = \ell^i_t$ for all i = 1, ..., N, or using π^i_t to denote orthogonal projection onto $\mathcal{I}^i_t$, $\pi^i_t \ell^i_t = \ell^i_t$. Taking the expectation E^i_t of (16) yields

$$\pi^i_t \ell^i_{t+1} = (1-\delta)\ell^i_t + \sigma\theta\{\pi^i_t \ell^i_{t+1} - \frac{1}{N}\pi^i_t(\Sigma_j \pi^j_t \ell^j_{t+1})\} + \pi^i_t(g^i_{t+1} - g^a_{t+1}),$$

where we use $(\pi^i_t)^2 = \pi^i_t$, or

$$\pi^i_t \ell^i_{t+1} = (1-\sigma\theta)^{-1}\{(1-\delta)\ell^i_t - \frac{\sigma\theta}{N}\pi^i_t(\Sigma_j \pi^j_t \ell^j_{t+1}) + \pi^i_t(g^i_{t+1} - g^a_{t+1})\}.$$

Summing over i we note that

(17) $$\Sigma_i \pi^i_t \ell^i_{t+1} = -(1-\sigma\theta)^{-1}\frac{\sigma\theta}{N}(\Sigma_i \pi^i_t)(\Sigma_j \pi^i_t \ell^i_{t+1}) + (1-\sigma\theta)^{-1}\Sigma_i \pi^i_t(g^i_{t+1} - g^a_{t+1})$$

where we used $\Sigma_j \ell^i = 0$.

Now we consider two cases in turn: A common information pattern and a differential information pattern.

<u>Case of Common Information Patten</u> Assume that $\mathcal{I}^i_t = \mathcal{I}_t$ hence the orthogonal projection operator π^i_t is the same as π_t for all i. In this case (17) reduces to a trivial relation "0 = 0". Equations (15) and (16) become

(18) $$L^a_{t+1} = (1 - \delta)L^a_t + g^a_{t+1},$$

and

$$\ell^i_{t+1} = (1 - \delta)\ell^i_t + \sigma\theta\pi_t\ell^i_{t+1} + g^i_{t+1} - g^a_{t+1},$$

or since $\pi_t\ell^i_{t+1} = (1 - \sigma\theta)^{-1}\{(1 - \delta)\ell^i_t + \pi_t(g^i_{t+1} - g^a_{t+1})\}$, the dynamics become*

(19) $$\ell^i_{t+1} = \kappa\ell^i_t + (g^i_{t+1} - g^a_{t+1}) + \frac{\sigma\theta}{1-\sigma\theta}\pi_t(g^i_{t+1} - g^a_{t+1}),$$

where

$$\kappa = (1 - \delta)/(1 - \sigma\theta).$$

We assume that $\kappa < 1$, i.e., that $\delta > \sigma\theta$. Suppose now that the exogenous disturbances on individual sectors are specified by

$$g^i_{t+1} = \{\varphi_0 + Lf_0(L)\}\varepsilon^0_{t+1} + \{\varphi_i + Lf_i(L)\}\varepsilon^i_{t+1}$$

where ε^0_t, ε^i_t, i=1, ..., N are the primitive independent random variables with zero means and unit variances.

Thus $g^i_{t+1} - g^a_{t+1} = \{\varphi_i + Lf_i(L)\}\varepsilon^i_{t+1} - \frac{1}{N}\Sigma_j\{\varphi_j + Lf_j(L)\}\varepsilon^j_{t+1}$. The information set $\mathcal{I}_t$ is the minimum closed subspace spanned by ε^0_t, ε^1_t, ..., ε^N_t and their past values.

Hence

$$\pi_t(g^i_{t+1} - g^a_{t+1}) = \pi_t\{\varphi_i + Lf_i(L)\}\varepsilon^i_{t+1} - \frac{1}{N}\Sigma_j\{\varphi_j + Lf_j(L)\}\varepsilon^j_{t+1}$$
$$= f_i(L)\varepsilon^i_t - \frac{1}{N}\Sigma_j f_j(L)\varepsilon^j_t.$$

Equation (19) thus becomes

$$(1 - L)\ell^i_{t+1} = \{\varphi_i + Lf_i(L)\}\varepsilon^i_{t+1} - \frac{1}{N}\Sigma_j\{\varphi_j + Lf_j(L)\}\varepsilon^j_{t+1}$$
$$+ \frac{\sigma\theta}{1-\sigma\theta}\{f_i(L)\varepsilon^i_t - \frac{1}{N}\Sigma_j f_j(L)\varepsilon^i_t\}.$$

Postulate that

(20) $$\ell^i_t = \Sigma_j\{\gamma^i_j + Lh^i_j(L)\}\varepsilon^j_t.$$

Then (20) satisfies (19) when we choose γ's and h's by

$$\gamma^i_j = \varphi_i(N - 1)/N,$$

---

* This type of dynamic equation is not considered by Gourieroux et al. [1979].

(21) $$h_i^i(L) = \frac{1}{1-\sigma\theta} f_i(L) + \kappa\varphi_i \frac{N-1}{N(1-\kappa L)},$$

$$\gamma_j^i = -\frac{1}{N}\varphi_j,\ h_j^i(L) = -\frac{1}{N(1-\kappa z)}\{\frac{1}{1-\sigma\theta}f_j(L) + \kappa\varphi_j\},\quad j \neq i$$

where we note that $1/\kappa$ is greater than one due to our assumption. Equation (21) shows that $h_j^i(z)$ is the same for all i. Hence we write h_j for h_j^i. The same for γ's. Note that the solution (20) exhibits serially correlated disturbances (business cycles), if and only if the exogenous disturbances are serially correlated.

From (18), the aggregate dynamics become

(18') $$L_{t+1}^a = (1-\delta)L_t^a + \{\varphi_0 + Lf_0(L)\}\varepsilon_{t+1}^0 + \frac{1}{N}\Sigma_j\{\varphi_j + Lf_j(L)\}\varepsilon_{t+1}^j.$$

Suppose that f_0 and f_j are all zero. Then the covariance sequence shows no serial correlation.

<u>Case of Differential Information Set</u> Suppose now that π_t^i is the orthogonal projection onto the subspace spanned by ε_t^0, ε_t^i and their lagged values. Let the exogenous disturbances be the same as in the previous case.

Suppose

$$\ell_{t+1}^i = \Sigma_j(\gamma_j^i + Lh_j(L))\varepsilon_{t+1}^j.$$

Then

$$\pi_t^i\ell_{t+1}^i = h_i^i(L)\varepsilon_t^i.$$

Hence

$$\Sigma_i\pi_t^i\ell_{t+1}^i = \Sigma_i h_i^i(L)\varepsilon_t^i.$$

Then

$$\pi_t^i(\Sigma_j\pi_t^j\ell_{t+1}^j) = h_i^i(L)\varepsilon_t^i.$$

Equation (16) becomes

$$\{1-(1-\delta)L\}\ell_{t+1}^i = \sigma\theta h_i^i(L)\varepsilon_t^i - \frac{\sigma\theta}{N}\Sigma_j h_j^i(L)\varepsilon_t^j$$

$$+ \{\varphi_i + Lf_i(L)\}\varepsilon_{t+1}^i - \frac{1}{N}\Sigma_j\{\varphi_j + Lf_j(L)\}\varepsilon_{t+1}^j,$$

or

$$\gamma_j^i = -\varphi_j/N$$

(22)

$$h_j^i(z) = -\frac{1}{N}\{(1-\delta)\varphi_j + f_i(z)\}/\{1-(1-\delta)z + \sigma\theta/N\},$$

$$\gamma_i^i = \frac{N-1}{N}\varphi_i,$$

and

$$h_i^i(z) = \frac{N-1}{N}\{(1-\delta)\varphi_i + f_i(z)\}/\{1-(1-\delta)z + \sigma\theta/N\}.$$

The root $z = (1+\sigma\theta/N)/(1-\delta)$ lies outside the unit disc, hence $h_j^i(z^{-1})$ is analytic inside $|z| < 1$, as required for the construction to be valid.

Because $\kappa < (1-\delta)/(1+\sigma\theta/N)$, comparison of (21) with (22) reveals that serial correlations in ℓ_t^i die out more slowly with the common information set in this simple example.

The aggregate dynamic equation (15) becomes

(15')
$$L_{t+1}^a = (1-\delta)L_t^a + \frac{\sigma\theta}{N}\Sigma_j h_j(L)\Sigma_t^j + \{\varphi_0 + Lf_0(L)\}\varepsilon_{t+1}^0 + \frac{1}{N}\Sigma_j\{\varphi_j + Lf_j(L)\}\varepsilon_{t+1}^j.$$

Compare (15') with (18'). We note that the aggregate dynamics under differential information pattern are more complex since the effects of Σ_t^j is not simply $\frac{1}{N}f_j(L)\varepsilon_t^j$ as in (18') but are given by

$$\frac{1}{N}f_j(L)\frac{1-(1-\delta)L}{1+\frac{\sigma\theta}{N}-(1-\delta)L}\varepsilon_t^j - \frac{\sigma\theta}{N}\frac{1-\delta}{N}\varphi_j\varepsilon_t^j.$$

As N becomes very large, the difference approaches zero, however.

Even when f_0 and f_1 are all zero, disturbances in (15') are now serially correlated. This is the most significant consequence of the differential information pattern.

14 NUMERICAL EXAMPLES

Two vector-valued time series from monthly observations on the Japanese economy have been used to estimate innovation models. The vector y_t is five-dimensional in one case and six-dimensional in the other, both covering the period of January 1975 to January 1982. All together there are 85 monthly observations per component. The five components are M_2 + CD (money supply outstanding, and quasi money plus certificates of deposit) in 0.1 Billion ¥; call rate (in Tokyo, unconditional, average; free rate after April 1979); Exchange rate, ¥/\$, (customs clearance-conversion rate, exports); production index-miniming and manufacturing (seasonally adjusted), and the wholesale price index-all commodities. The base year for these two indices is 1975. The sixth component for the second time series records the current account in million dollars. The five data series are plotted in Figures 1 ∿ 5. They are tabulated in Table 1. The three series for M_2 + CD, WPI and Index of production are further processed by taking the first difference of their respective logarithms. They are shown in Figures 6 ∿ 8, where L stand for first difference of the logarithms. We note seemingly random scatters rather than trend growths that are visible in the original data. They are tabulated in Table 2.* The data are further transformed by subtracting sample means and dividing by the sample standard deviation to produce the two time series $\{y_t\}$ which are both mean zero, and of full rank.

The first 70 data points were used to fit an AR model to the five-dimensional series by an AIC program supplied by Dr. H. Akaike. The program produced AR(2)

$$y_t = B(1)y_{t-1} + B(2)y_{t-2} + x_t$$

where the 5 × 5 matrice B(i), i = 1, 2 and the covariance of x_t are printed out in Table 3.

---

* Because of differencing, the maximum usable data points are 84.

Table 1

	KO	EX	CALL	M	WPI	CUA
1	98.8000	300.890	12.6740	.106833E+07	100.400	-1164.00
2	97.9000	297.100	13.0000	.107348E+07	99.7000	109.000
3	96.6000	287.920	12.9200	.109375E+07	99.3000	130.000
4	98.9000	290.570	12.0200	.111028E+07	99.5000	185.000
5	98.9000	291.940	11.0600	.111588E+07	99.5000	-574.000
6	99.9000	291.840	10.7200	.113824E+07	99.3000	89.0000
7	101.000	295.610	11.0000	.114483E+07	99.4000	-4.00000
8	100.800	297.310	10.6920	.115610E+07	99.9000	22.0000
9	101.700	298.100	9.66700	.116459E+07	100.200	-41.0000
10	102.100	302.380	8.73100	.117207E+07	100.700	-156.000
11	100.400	302.040	7.60900	.120570E+07	100.700	37.0000
12	102.600	304.750	7.96300	.125330E+07	101.300	685.000
13	104.400	305.470	7.28300	.123065E+07	102.100	-1081.00
14	106.900	302.720	7.00000	.124883E+07	102.800	147.000
15	108.600	301.450	7.00000	.126235E+07	103.300	825.000
16	110.000	299.280	6.75000	.128177E+07	103.900	292.000
17	109.600	299.010	6.75000	.129405E+07	104.200	226.000
18	111.700	299.840	6.90400	.132193E+07	104.700	423.000
19	112.700	296.840	7.08300	.132389E+07	105.600	410.000
20	113.000	292.760	7.25000	.132379E+07	106.100	13.0000
21	112.800	288.170	7.05200	.134482E+07	106.500	560.000
22	112.800	288.530	6.77000	.135556E+07	106.800	637.000
23	114.500	294.100	6.77100	.137034E+07	107.100	40.0000
24	115.100	295.680	7.11100	.142249E+07	107.300	1188.00
25	115.900	292.450	7.00000	.139133E+07	107.200	-650.000
26	114.300	288.270	7.00000	.139423E+07	107.500	683.000
27	116.000	282.440	6.69200	.142350E+07	107.500	860.000
28	115.300	275.720	5.87000	.143041E+07	107.500	1226.00
29	114.800	277.660	5.18200	.143960E+07	107.700	85.0000
30	115.700	275.640	5.47600	.147144E+07	107.300	872.000
31	113.500	267.590	5.65900	.148676E+07	106.800	1494.00
32	116.000	265.710	5.75000	.147124E+07	107.000	669.000
33	115.800	267.120	4.97900	.148910E+07	107.100	1098.00
34	115.000	261.590	4.91500	.148856E+07	106.800	1316.00
35	117.300	249.170	4.62000	.151905E+07	106.100	1111.00
36	118.100	241.690	5.01400	.158033E+07	105.700	2154.00
37	118.700	240.800	4.78800	.154040E+07	105.600	-266.000
38	119.400	241.440	4.80400	.154600E+07	105.700	1835.00
39	120.700	236.630	4.62000	.157332E+07	105.600	2402.00
40	121.400	222.970	4.14100	.161168E+07	105.200	1680.00
40	121.400	222.970	4.14100	.161168E+07	105.200	1680.00
41	122.000	225.390	4.06000	.161041E+07	105.500	634.000
42	122.200	222.710	4.10600	.165076E+07	105.100	2265.00
43	122.100	205.270	4.44200	.165489E+07	104.100	1989.00
44	123.600	190.940	4.39400	.165349E+07	103.200	1246.00
45	124.700	190.920	4.25000	.167462E+07	103.100	1911.00
46	125.400	187.700	4.18000	.167206E+07	102.500	393.000
47	125.900	184.890	3.93200	.170669E+07	102.700	592.000
48	127.300	196.530	4.56700	.178720E+07	103.300	1853.00
49	127.700	196.240	4.28800	.172604E+07	103.900	-1462.00
50	129.300	199.150	4.34800	.173261E+07	104.800	262.000
51	129.100	203.630	4.63900	.177588E+07	105.700	489.000
52	130.000	211.310	4.88540	.181619E+07	107.500	-345.000
53	132.200	217.540	5.11500	.180470E+07	109.200	-889.000
54	132.500	219.790	5.34380	.184497E+07	110.600	108.000
55	134.200	217.250	5.80290	.184227E+07	112.700	-939.000
56	134.700	216.160	6.68520	.184383E+07	114.500	-1510.00
57	133.600	220.590	6.80980	.187794E+07	116.100	-780.000
58	136.400	225.520	6.74280	.185546E+07	117.400	-1086.00
59	138.200	238.650	7.58070	.188678E+07	119.200	-2294.00
60	138.500	243.760	8.04570	.195013E+07	121.400	-308.000
61	140.100	237.400	8.05710	.190033E+07	124.000	-3372.00
62	146.100	240.460	8.73960	.190956E+07	127.200	-1250.00
63	142.700	247.450	10.7300	.194735E+07	129.800	-1188.00
64	144.500	252.520	12.2100	.198030E+07	133.300	-1784.00
65	143.000	237.750	12.5625	.196899E+07	133.100	-1861.00
66	142.400	221.170	12.6425	.200250E+07	133.000	-888.000
67	142.600	217.950	12.7014	.198973E+07	133.500	-951.000
68	137.200	224.660	12.0865	.200767E+07	134.500	-913.000
69	141.800	218.730	11.4036	.199238E+07	134.100	853.000
70	143.100	210.240	11.0361	.198972E+07	133.100	-17.0000
71	141.300	211.280	9.50000	.204781E+07	133.200	-506.000
72	143.500	212.420	9.48840	.208986E+07	133.000	1131.00
73	143.800	203.500	8.90760	.203756E+07	132.300	-2724.00
74	143.900	203.500	8.60330	.205005E+07	132.100	-129.000
75	144.200	207.760	8.03500	.208097E+07	132.100	777.000
76	144.700	212.250	7.18500	.212114E+07	132.700	449.000
77	143.000	217.510	7.05730	.216326E+07	133.800	-382.000
78	146.300	223.990	7.11780	.217792E+07	134.400	1388.00
79	147.600	226.620	7.25930	.217770E+07	135.000	940.000
80	146.300	236.070	7.23560	.217886E+07	135.700	477.000
81	149.400	230.270	7.25780	.219201E+07	135.700	2114.00
82	151.000	229.050	7.05050	.220312E+07	135.500	1788.00
83	150.900	230.150	6.79890	.224106E+07	135.300	-1061.00
84	150.200	218.180	6.70140	.232042E+07	135.100	1133.00
85	149.300	221.650	6.57610	.228666E+07	135.100	-1892.00
	1	2	3	4	5	6

Table 2

	LKO	LM	LW
2	-.915099E-02	.481369E-02	-.699650E-02
3	-.133678E-01	.187009E-01	-.402020E-02
4	.235305E-01	.149992E-01	.201220E-02
5	.0	.503559E-02	.0
6	.100604E-01	.198337E-01	-.201220E-02
7	.109509E-01	.577297E-02	.100660E-02
8	-.198228E-02	.979962E-02	.501757E-02
9	.888904E-02	.731683E-02	.299853E-02
10	.392536E-02	.639977E-02	.497761E-02
11	-.167905E-01	.282948E-01	.0
12	.216757E-01	.387205E-01	.594052E-02
13	.173918E-01	-.182448E-01	.786634E-02
14	.236641E-01	.146687E-01	.683260E-02
15	.157776E-01	.107656E-01	.485202E-02
16	.128090E-01	.152677E-01	.579158E-02
17	-.364307E-02	.953723E-02	.288326E-02
18	.189794E-01	.213122E-01	.478699E-02
19	.891271E-02	.148536E-02	.855919E-02
20	.265842E-02	-.778040E-04	.472367E-02
21	-.177159E-02	.157591E-01	.376303E-02
22	.0	.795598E-02	.281283E-02
23	.149586E-01	.108472E-01	.280508E-02
24	.522641E-02	.373471E-01	.186564E-02
25	.692646E-02	-.221495E-01	-.932316E-03
26	-.139012E-01	.208361E-02	.279463E-02
27	.147637E-01	.207778E-01	.0
28	-.605287E-02	.483899E-02	.0
29	-.434594E-02	.640559E-02	.185871E-02
30	.780923E-02	.218776E-01	-.372102E-02
31	-.191978E-01	.103557E-01	-.467072E-02
32	.217874E-01	-.104883E-01	.187102E-02
33	-.172573E-02	.120649E-01	.934057E-03
34	-.693233E-02	-.366731E-03	-.280508E-02
35	.198025E-01	.202760E-01	-.657585E-02
36	.679699E-02	.395518E-01	-.377710E-02
37	.506763E-02	-.255922E-01	-.946579E-03
38	.587987E-02	.362689E-02	.946579E-03
39	.108289E-01	.175178E-01	-.946579E-03
40	.578272E-02	.240910E-01	-.379501E-02
41	.493021E-02	-.787687E-03	.284768E-02
42	.163798E-02	.247463E-01	-.379876E-02
43	-.818715E-03	.250056E-02	-.956030E-02
44	.122102E-01	-.848148E-03	-.868306E-02
45	.886036E-02	.126957E-01	-.969521E-03
46	.559775E-02	-.152928E-02	-.583650E-02
47	.397931E-02	.205006E-01	.194929E-02
48	.110585E-01	.460949E-01	.582517E-02
49	.313733E-02	-.348221E-01	.579158E-02
50	.124515E-01	.380033E-02	.862481E-02
51	-.154796E-02	.246643E-01	.855121E-02
52	.694722E-02	.224449E-01	.168860E-01
53	.167815E-01	-.634488E-02	.156902E-01
54	.226674E-02	.220698E-01	.127390E-01
55	.127486E-01	-.146342E-02	.188094E-01
56	.371886E-02	.844795E-03	.158454E-01
57	-.819987E-02	.183326E-01	.138770E-01
58	.207415E-01	-.120465E-01	.111350E-01
59	.131102E-01	.167428E-01	.152159E-01
60	.216844E-02	.330223E-01	.182881E-01
61	.114861E-01	-.258680E-01	.211907E-01
62	.419349E-01	.484738E-02	.254791E-01
63	-.235468E-01	.195940E-01	.202341E-01
64	.125350E-01	.167804E-01	.266074E-01
65	-.104349E-01	-.572661E-02	-.150148E-02
66	-.420467E-02	.168756E-01	-.751528E-03
67	.140349E-02	-.639894E-02	.375235E-02
68	-.386037E-01	.897589E-02	.746272E-02
69	.329778E-01	-.764544E-02	-.297848E-02
70	.912609E-02	-.133648E-02	-.748506E-02
71	-.126584E-01	.287770E-01	.751078E-03
72	.154498E-01	.203262E-01	-.150261E-02
	1	2	3

Table 3

MATRIX-B (1).

MATRIX 5 X 5

	1	2	3	4	5
1	-0.11733D+01	0.26499D+01	-0.31725D+02	-0.13199D+03	0.92415D+02
2	-0.35550D-02	-0.11664D+01	0.36755D+01	-0.33368D+01	-0.27138D+01
3	-0.24934D-03	0.21824D-02	0.47058D+00	0.39212D-01	-0.22154D-01
4	0.46405D-04	0.38607D-02	-0.76165D-01	0.60926D+00	-0.27524D+00
5	-0.31625D-03	0.91974D-03	-0.12382D+00	0.45711D+00	0.16445D-01

MATRIX-B (2).

MATRIX 5 X 5

	1	2	3	4	5
1	0.22445D+00	-0.32331D+01	0.91775D+01	-0.16796D+03	0.51880D+02
2	0.48969D-02	0.23043D+00	0.15188D+01	0.34976D+01	-0.93840D+01
3	0.18649D-03	-0.15371D-02	0.46790D+00	-0.52973D-01	-0.48617D-01
4	-0.55651D-04	-0.28898D-02	-0.14196D+00	0.24543D+00	-0.25477D+00
5	0.28290D-03	0.37448D-03	-0.23494D+00	0.23501D+00	-0.41859D-01

MATRIX S

MATRIX 5 X 5

	1	2	3	4	5
1	0.76064D+02	0.33648D+01	0.13081D-01	0.26044D-02	0.28933D-01
2	0.33648D+01	0.73099D+00	0.26507D-03	-0.37685D-03	0.11730D-02
3	0.13081D-01	0.26507D-03	0.16339D-03	-0.13389D-04	-0.19593D-04
4	0.26044D-02	-0.37685D-03	-0.13389D-04	0.11355D-03	0.97673D-04
5	0.28933D-01	0.11730D-02	-0.19593D-04	0.97673D-04	0.22893D-03

Figures 9 $\sim$ 13 plots the predicted values and actual values. The symbol ▲ denotes (long-range) predicted value with no observation, while the letter x denotes predicted values for which observation marked by ● is available.

The data $\{y_t\}$, t=1, ..., N, are also used to calculate sample covariance matrices Λ_ℓ, calculated by $(\Sigma_{t=1}^{N-\ell} y_{t+\ell}y_t')/N$, to construct the Hankel matrix. Its singular values and the eigenvectors to construct U and V, needed is the singular value decomposition, are also calculated. By inspection of the singular values, n = 10 seems to be the largest useful dimension of approximate innovation models and the matrices A, C, and M were calculated accordingly following the procedure of Section 9.2. Visual inspection of the singular values reveals that n = 1, 2 and 6 could also be possible dimensions of the state space models.

As examples, we choose n to be 2 and 3 with N = 74 and 84 and numerically solve the algebraic Riccati equation to construct an approximate innovation model

$$z_{t+1} = Az_t + \Gamma e_t,$$

$$y_t = Cz_t + e_t,$$

where

$$Z = Ez_t z_t',$$

$$Z = AZA' + (M - AZC')(\Lambda_0 - CZC')^{-1}(M - AZC')',$$

$$K = M - AZC',$$

$$\Sigma = \Lambda_0 - CZC',$$

$$\Gamma = K\Sigma^{-1},$$

following the procedure of Section 10.3. The model is then used to predict out-of-sample y's by first calculating $z_{N+1|N}$ from $z_{t+1|t} = (A - \Gamma C)z_{t|t-1} + \Gamma y_t$, $z_{0|-1} = 0$, t = 0, ..., N. Then the predicted values are generated by $y_{N+k|N} = Cz_{N+k|N} = CA^{k-1}z_{N+1|N}$, k = 1, 2, ...

The covariance matrix Λ_0 with 74 data points is shown below

$$\Lambda_0 = \begin{pmatrix} 1.00 & 0.02 & -0.16 & -0.09 & 0.17 \\ & 1.00 & 0.27 & 0.09 & -0.03 \\ & & 1.00 & -0.03 & 0.06 \\ & & & 1.00 & 0.01 \\ & & & & 1.00 \end{pmatrix}.$$

With 84 data points, the matrix changes into

$$\Lambda_0 = \begin{pmatrix} 1.00 & 0.03 & -0.15 & -0.10 & 0.16 \\ & 1.00 & 0.27 & 0.08 & -0.02 \\ & & 1.00 & -0.03 & 0.06 \\ & & & 1.00 & 0.01 \\ & & & & 1.00 \end{pmatrix}.$$

The elements vary in the second places below the decimal points, showing that the number of observations is too small for the matrix elements to have converged. In the first example, N is 74, i.e., the data from January 1975 to March 1981 have been used to construct the Hankel matrix. The largest reduction in its singluar value occur from σ_1, to σ_2 the next largest drop arise going from σ_2 to σ_3. The dimension of the approximate model is taken to be two. The matrices A, C. M are calculated to be

$$A = \begin{pmatrix} -0.15 & 0.18 \\ 0.02 & 0.12 \end{pmatrix},$$

$$M = \begin{pmatrix} -0.14 & 0.14 & -0.03 & -1.08 & -0.55 \\ 0.42 & 0.60 & 2.14 & 0.96 & 0.30 \end{pmatrix},$$

$$C = \begin{pmatrix} 0.01 & -0.10 \\ -0.86 & 0.10 \\ -0.23 & 0.91 \\ -0.06 & -0.08 \\ 0.47 & 0.36 \end{pmatrix}.$$

The solution of the Ricatti equation is:

$$Z = \begin{pmatrix} 1.62 & 1.30 \\ & 1.93 \end{pmatrix}.$$

In the second example, N is taken to be 84, i.e., the data from January 1975 to January 1982 have been used also to construct the two-diemnsional innovation model. Its matrices are

$$A = \begin{pmatrix} -0.15 & 0.18 \\ 0.04 & 0.16 \end{pmatrix},$$

$$M = \begin{pmatrix} -0.16 & 0.06 & -0.09 & -0.99 & -0.51 \\ 0.38 & 0.53 & 2.35 & 0.88 & 0.25 \end{pmatrix},$$

$$C = \begin{pmatrix} 0.00 & -0.09 \\ -0.88 & 0.16 \\ -0.18 & 0.92 \\ -0.07 & -0.07 \\ 0.45 & 0.33 \end{pmatrix}.$$

The Z matrix becomes

$$Z = \begin{pmatrix} 1.41 & -1.14 \\ & 2.56 \end{pmatrix}.$$

When the current account is added as the sixth element, its correlation with the other five elements are:

$$\begin{pmatrix} 0.03 \\ 0.04 \\ -0.46 \\ 0.47 \\ -0.57 \end{pmatrix}.$$

Using the data January 1975 to March 1981, we construct the three-dimensional innovation model as the third example:

$$A = \begin{pmatrix} 0.07 & -0.07 & -0.06 \\ 0.03 & 0.06 & 0.09 \\ 0.44 & 0.11 & 0.16 \end{pmatrix},$$

$$M = \begin{pmatrix} 0.33 & 0.02 & -0.38 & 0.34 & 1.01 & 0.56 \\ -0.24 & -0.22 & -0.78 & -0.52 & -3.38 & -1.20 \\ 0.08 & 0.09 & -2.51 & -0.07 & -5.63 & -2.81 \end{pmatrix},$$

$$C = \begin{pmatrix} -0.03 & -0.10 & -0.11 \\ -0.59 & 0.55 & -0.40 \\ 0.30 & 0.96 & 0.32 \\ -0.08 & -0.03 & -0.08 \\ 0.58 & 0.08 & -0.38 \\ -0.59 & -0.47 & 0.00 \end{pmatrix}.$$

The Z matrix is given by

$$Z = \begin{pmatrix} 3.06 & -11.95 & -25.28 \\ & 49.52 & 105.72 \\ & & 228.96 \end{pmatrix}.$$

The two-dimensional innovation model with N = 74 has, by construction, the (2×2) upper left hand corner of A, the first two rows of M, and by the first two columns of C. Its Z matrix becomes $Z = \begin{pmatrix} 3.50 & -12.72 \\ & 51.03 \end{pmatrix}$. This is the fourth example. The fifth model uses N = 84 with the five-dimensional vector time series.

These models have been used to calculate the next 10 data points. The models based on data of January 1975 to March 1981, thus calculate values for the period of April 81 through January 1982. The models based on the data of January 1975 through January 1982 predict values for the period of February 1982 through November 1982.

The models track trend growth paths of the Index of production, money stock and WPI fairly well. The predictions settle on neraly constant values for the exchange rate, the call rate and for the current account in the third and the fourth model. With 74 data points, the three-dimensional model with the six-dimensional data vector gives the smallest predicted value for the Index of production and WPI. The predictions of the two-dimensional model with the six-dimensional data vector lie in the middle, the highest predictions being generated by the two-dimensional model with the five-dimensinal data vector. The latter two models' predicted money stock nearly coincide. The first model gives the lowest predicted money stock. Numerical values are gives in Tables 4 and 5.

Table 4

Lowest predictions (dim y = 6, 3-dim model)

	Production Index %	$M \times 10^{-4}$	WPI %
4/81	144.8	213.3	131.4
5/81	145.5	215.3	131.7
6/81	146.3	217.3	132.2
7/81	147.0	219.3	132.7
8/81	147.8	221.3	133.2
9/81	148.5	223.4	133.7
10/81	149.3	225.4	134.2
11/81	150.0	227.5	134.7
12/81	150.8	229.6	135.2
1/82	151.6	231.7	135.7

Table 5

Highest predictions (dim y = 6, 2-dim model)

	Production Index %	$M\times10^{-4}$	WPI %
4/81	145.3	214.1	133.0
5/81	146.0	216.1	133.4
6/81	146.7	218.1	133.9
7/81	147.5	220.1	134.4
8/81	148.2	220.1	134.9
9/81	149.0	224.1	135.4
10/81	149.8	226.2	135.9
11/81	150.5	228.3	136.4
12/81	151.3	230.4	136.9
1/82	152.1	232.5	137.4

Predictions of the exchange rate and the call rate settle down to 250.4¥/$ and 7.38% quickly in all thc models. The current account seems to settle on the level 139.8M$. Table 6 list predicted values models.

Table 6

N	dim y	dim z	EX	CALL	CUA
74	5	2	250.4	7.38	____
74	6	3	250.4	7.38	139.8
74	6	2	250.4	7.38	139.8
84	6	2	247.8	7.34	178.4
84	5	2	247.8	9.34	139.0

The values of C_{32} ranges from 0.85 to 0.96. The element C_{21} in magnitude is about 0.86 $\sim$ 0.88 for the two-dimensional model and is about 0.6 for the three-dimensional model. These figures indicate that the exchange rate and the call rate or something close to them are being picked as the two state vector components in the two-diemnsional models. When the current account is added, the exchange rate's influence on the first state vector component diminishes.

TIME SERIES PLOT

Fig. 1

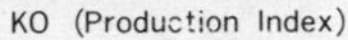

MINIMUM
96.6000

MAXIMUM
151.000

75– I
76– I
77– I
78– I
79– I
80– I
81– I
82– I

98.8000
97.9000
96.6000
98.9000
98.9000
99.9000
101.000
100.800
101.700
102.100
100.400
102.600
104.400
106.900
108.600
110.000
109.600
111.700
112.700
113.000
112.300
112.300
114.500
115.100
115.900
114.300
116.000
115.300
114.800
115.700
113.500
116.000
115.800
115.000
117.300
118.100
118.700
119.400
120.700
121.400
122.000
122.200
122.100
123.600
124.700
125.400
125.900
127.300
127.700
129.300
129.100
130.000
132.200
132.500
134.200
134.700
133.600
136.400
138.200
138.500
140.100
146.100
142.700
144.500
143.000
142.400
142.600
137.200
141.800
143.100
141.300
143.500
143.800
143.900
144.200
144.700
143.000
146.300
147.600
146.300
149.400
151.000
150.900
150.200
149.300

96.6000
MINIMUM

151.000
MAXIMUM

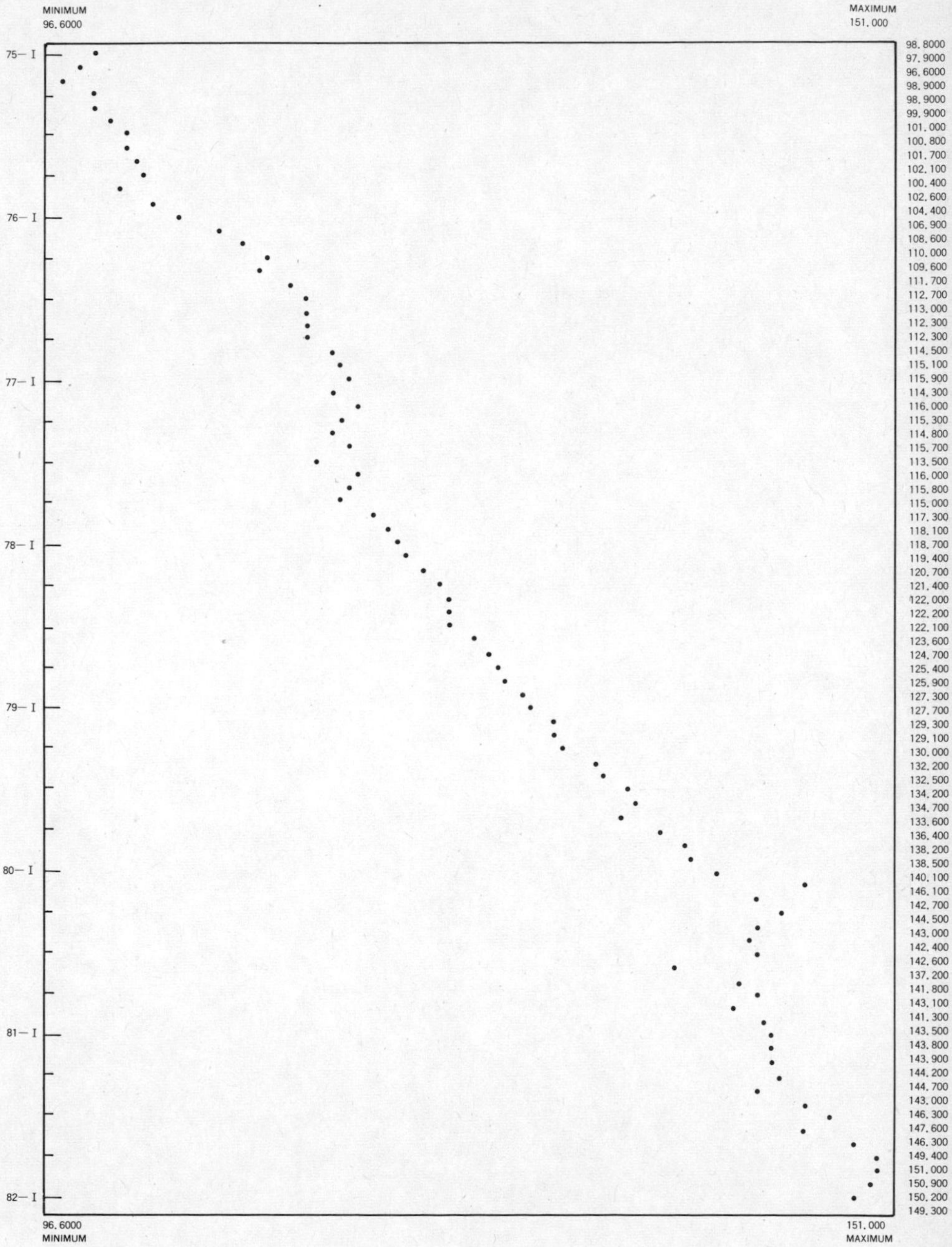

TIME SERIES PLOT

Fig. 2

Ex (Exchange Rate)

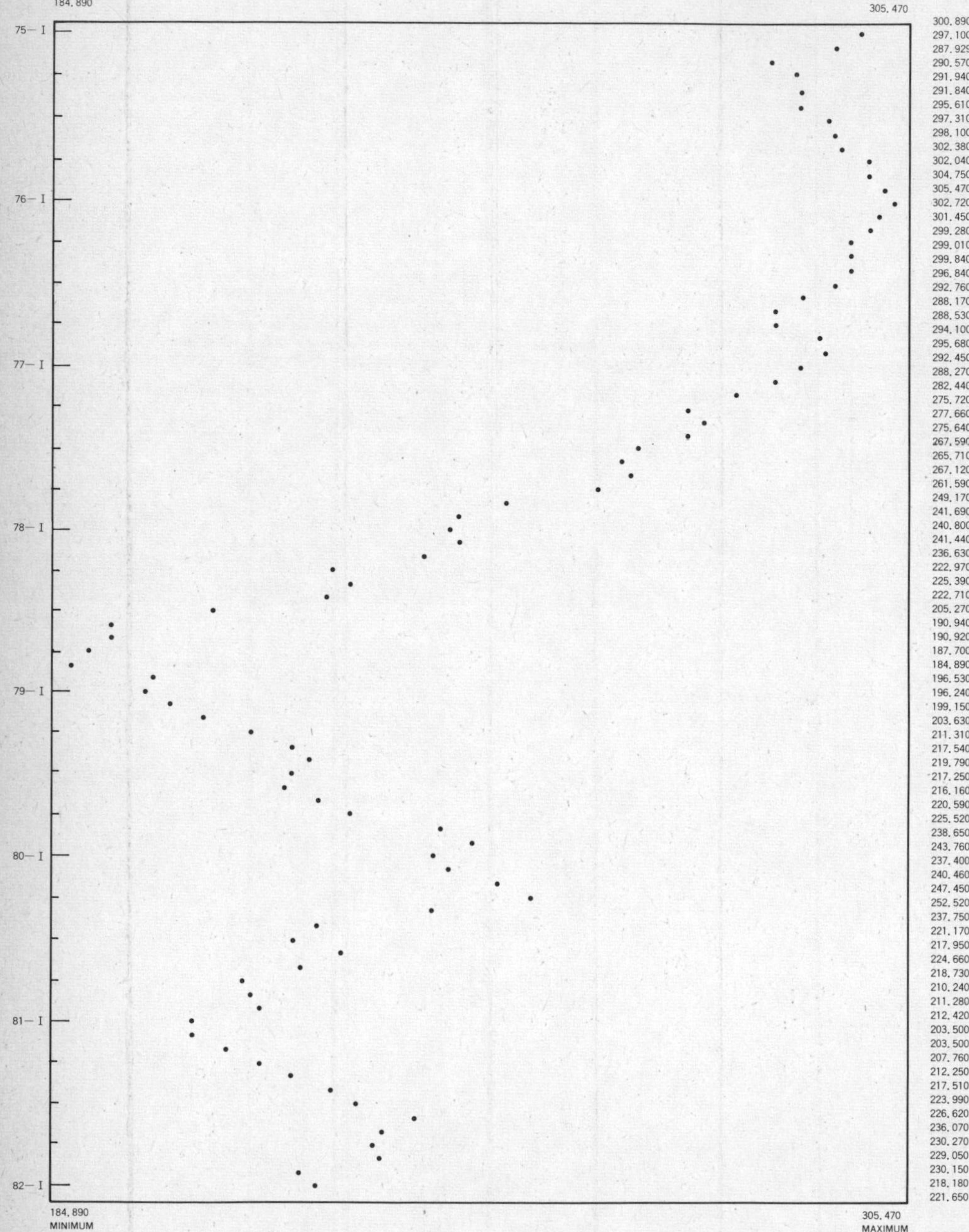

TIME SERIES PLOT

Fig. 3

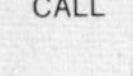

MINIMUM
3, 93200

MAXIMUM
13, 0000

75— I
76— I
77— I
78— I
79— I
80— I
81— I
82— I

12, 6740
13, 0000
12, 9200
12, 0200
11, 0600
10, 7200
11, 0000
10, 6900
9, 66700
8, 73100
7, 60900
7, 96300
7, 28300
7, 00000
7, 00000
6, 75000
6, 75000
6, 90400
7, 08300
7, 25000
7, 05200
6, 77000
6, 77100
7, 11100
7, 00000
7, 00000
6, 69200
5, 87000
5, 18200
5, 47600
5, 65900
5, 75000
4, 97900
4, 91500
4, 62000
5, 01400
4, 78800
4, 80400
4, 62000
4, 14100
4, 06000
4, 10600
4, 44200
4, 39400
4, 25000
4, 18000
3, 93200
4, 56700
4, 28800
4, 34800
4, 63900
4, 88540
5, 11500
5, 34380
5, 80290
6, 68520
6, 80980
6, 74280
7, 58070
8, 04570
8, 05710
8, 73960
10, 7300
12, 2100
12, 5625
12, 6425
12, 7014
12, 0865
11, 4036
11, 0361
9, 50000
9, 48840
8, 90760
8, 60330
8, 03500
7, 18500
7, 05730
7, 11780
7, 25930
7, 23560
7, 25780
7, 05050
6, 79890
6, 70140
6, 57610

3, 93200
MINIMUM

13, 0000
MAXIMUM

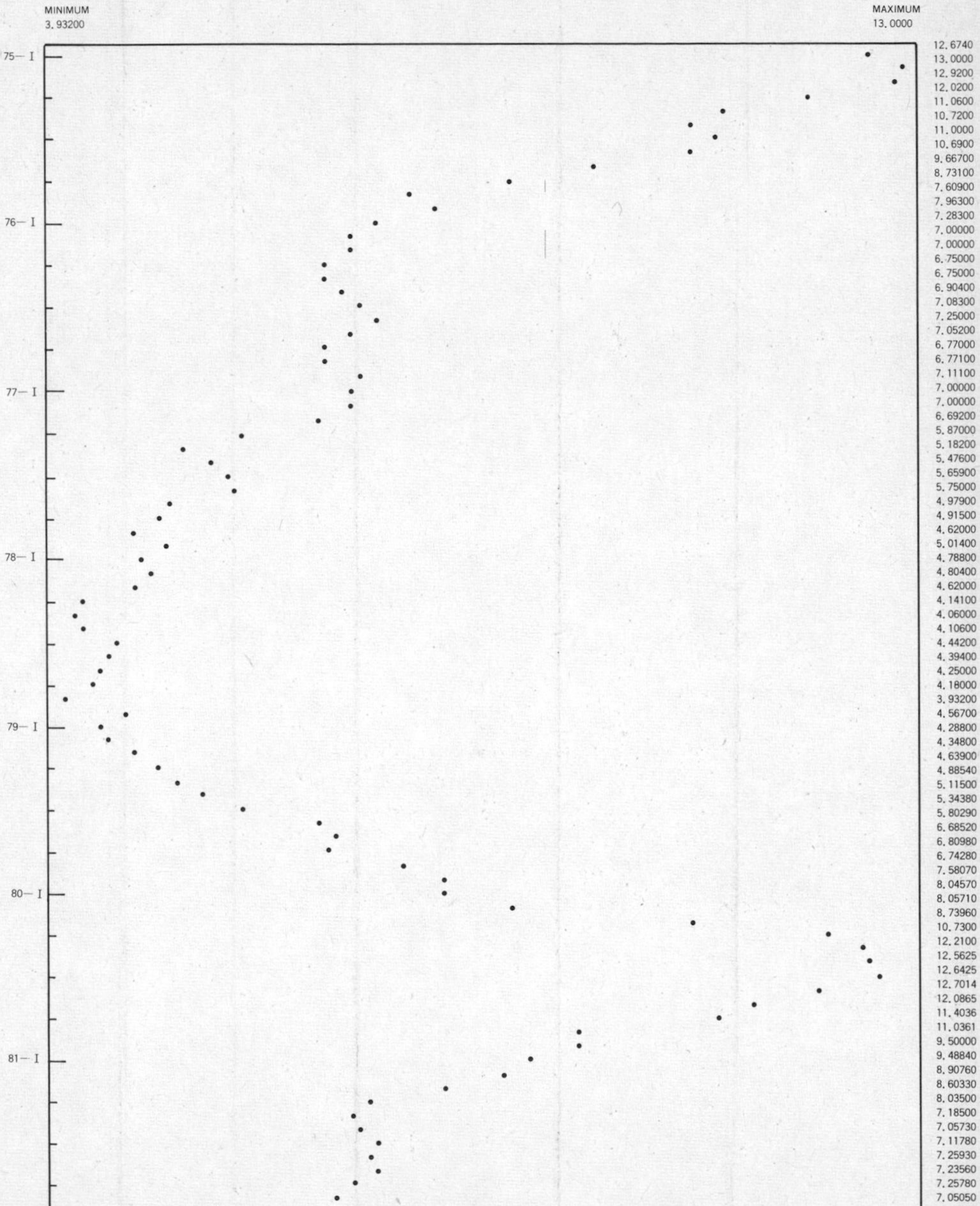

TIME SERIES PLOT

Fig. 4

M_2+CD

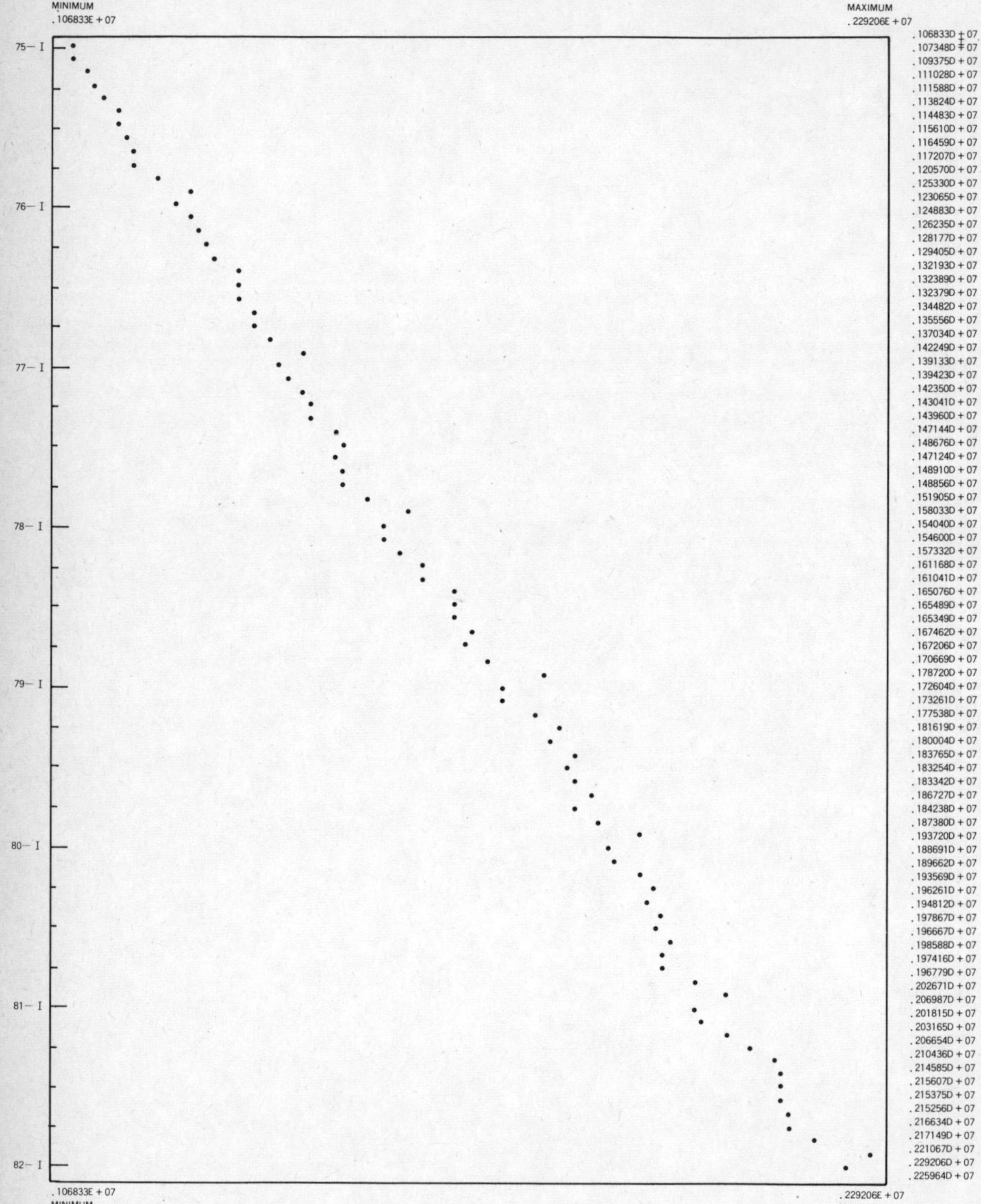

TIME SERIES PLOT

Fig. 5

WPI

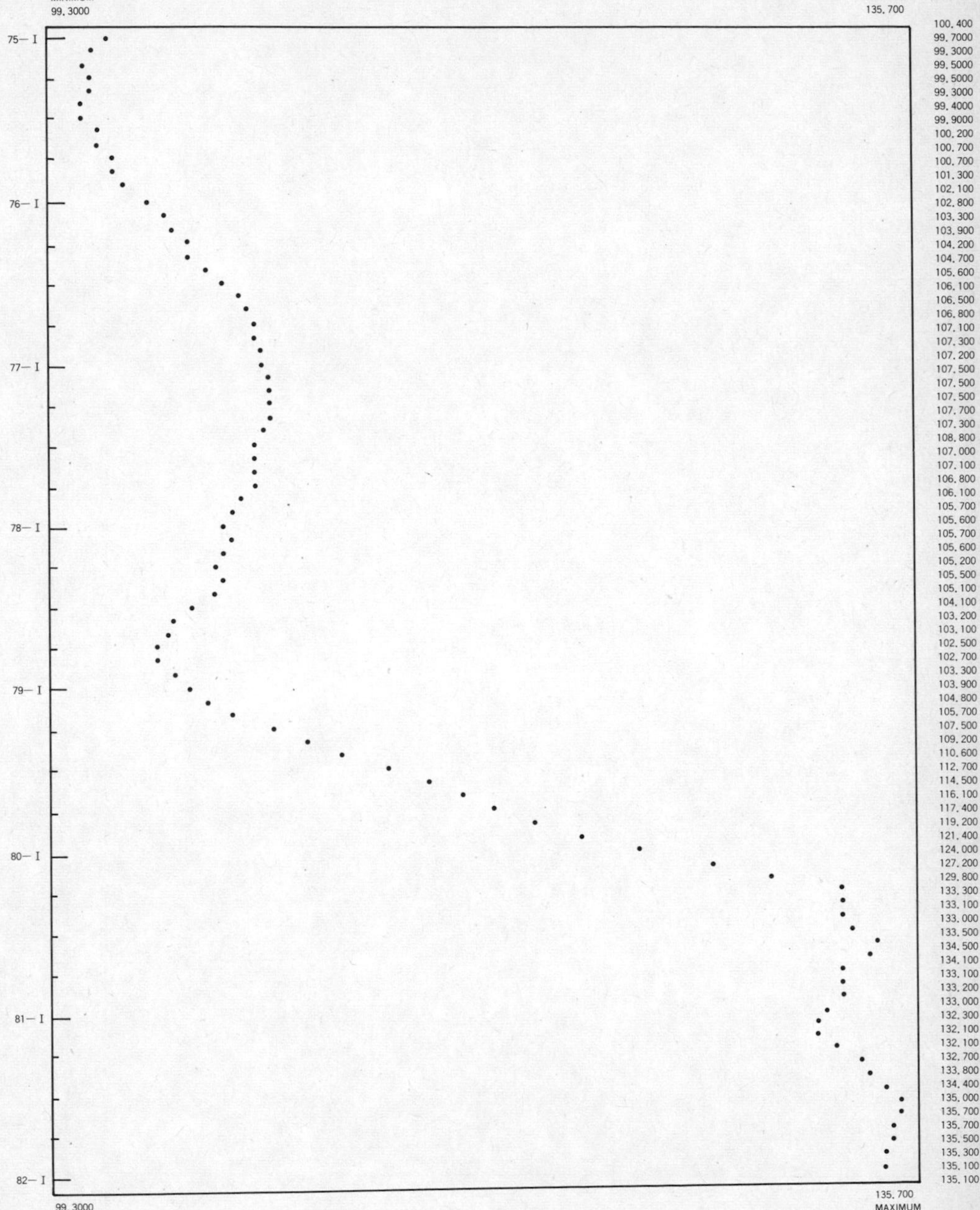

TIME SERIES PLOT

Fig. 6

LKO

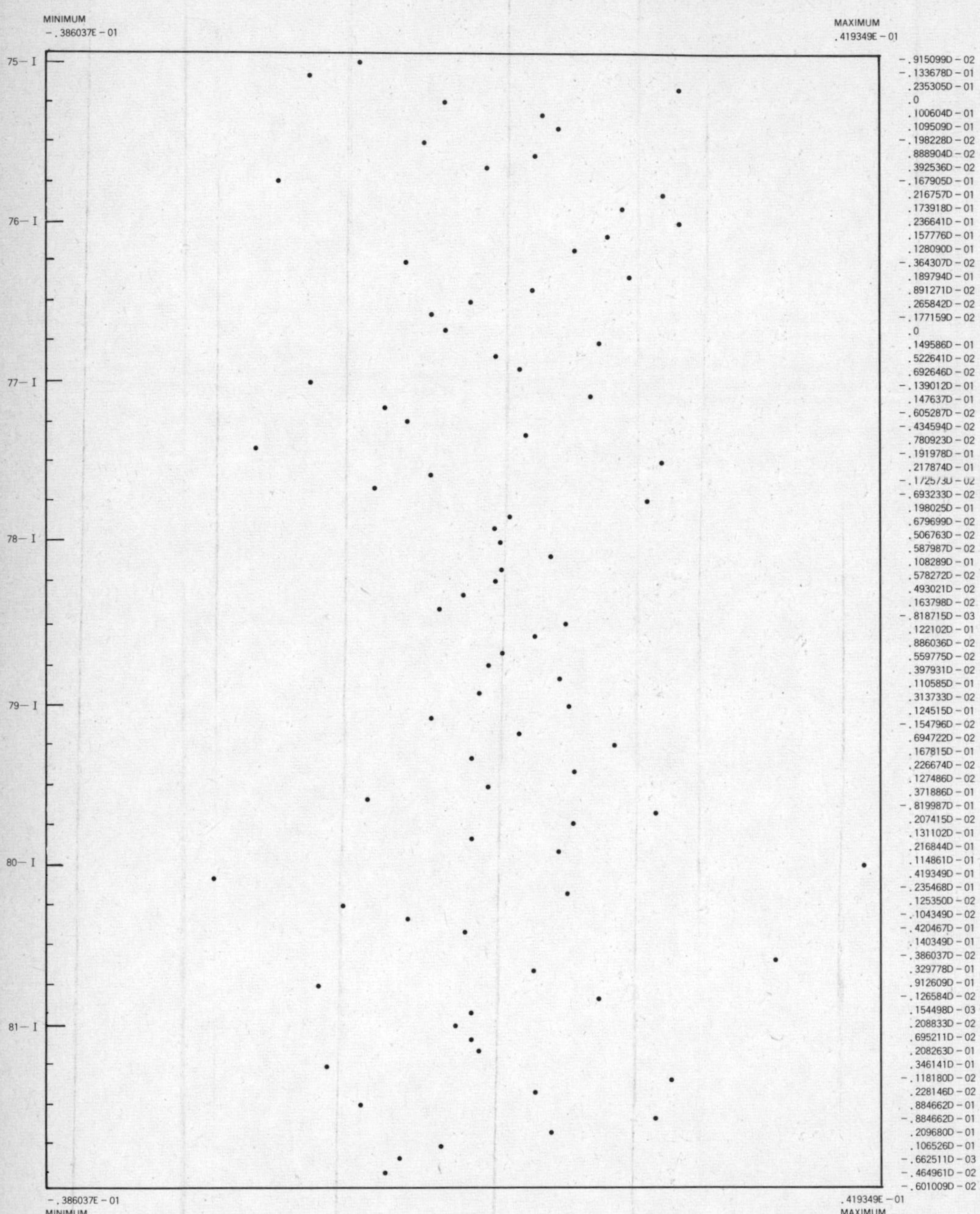

TIME SERIES PLOT

Fig. 7

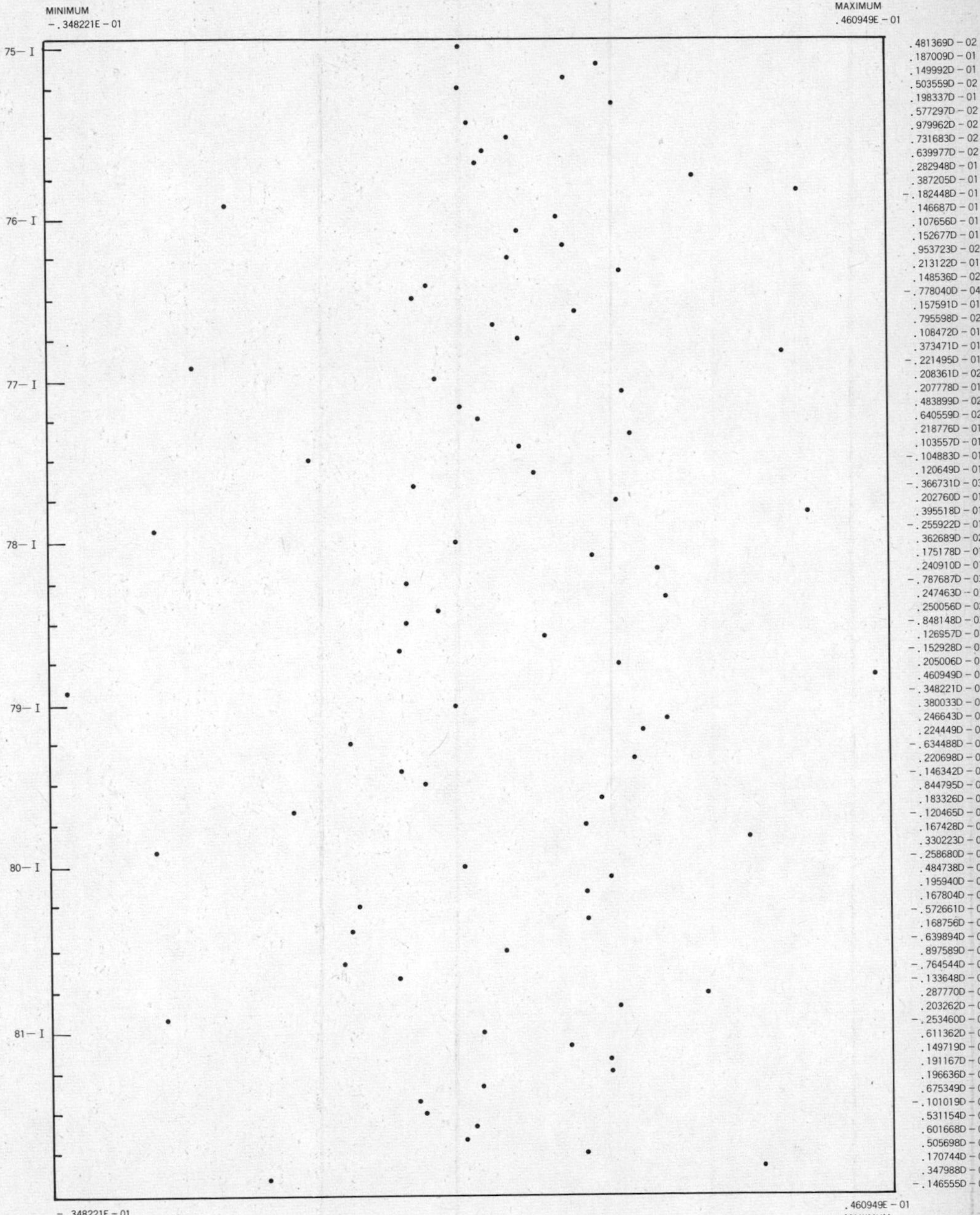

TIME SERIES PLOT

Fig. 8

LW

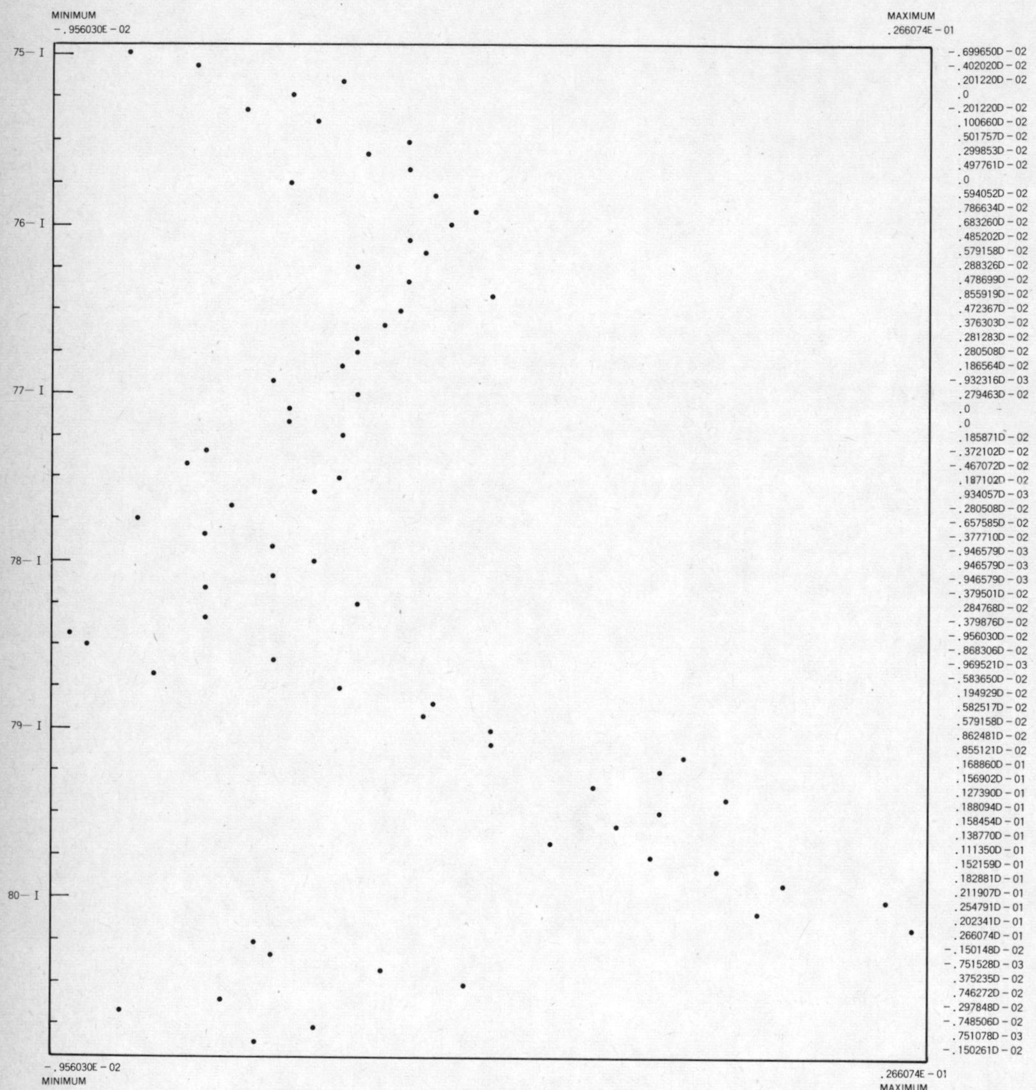

Fig. 9

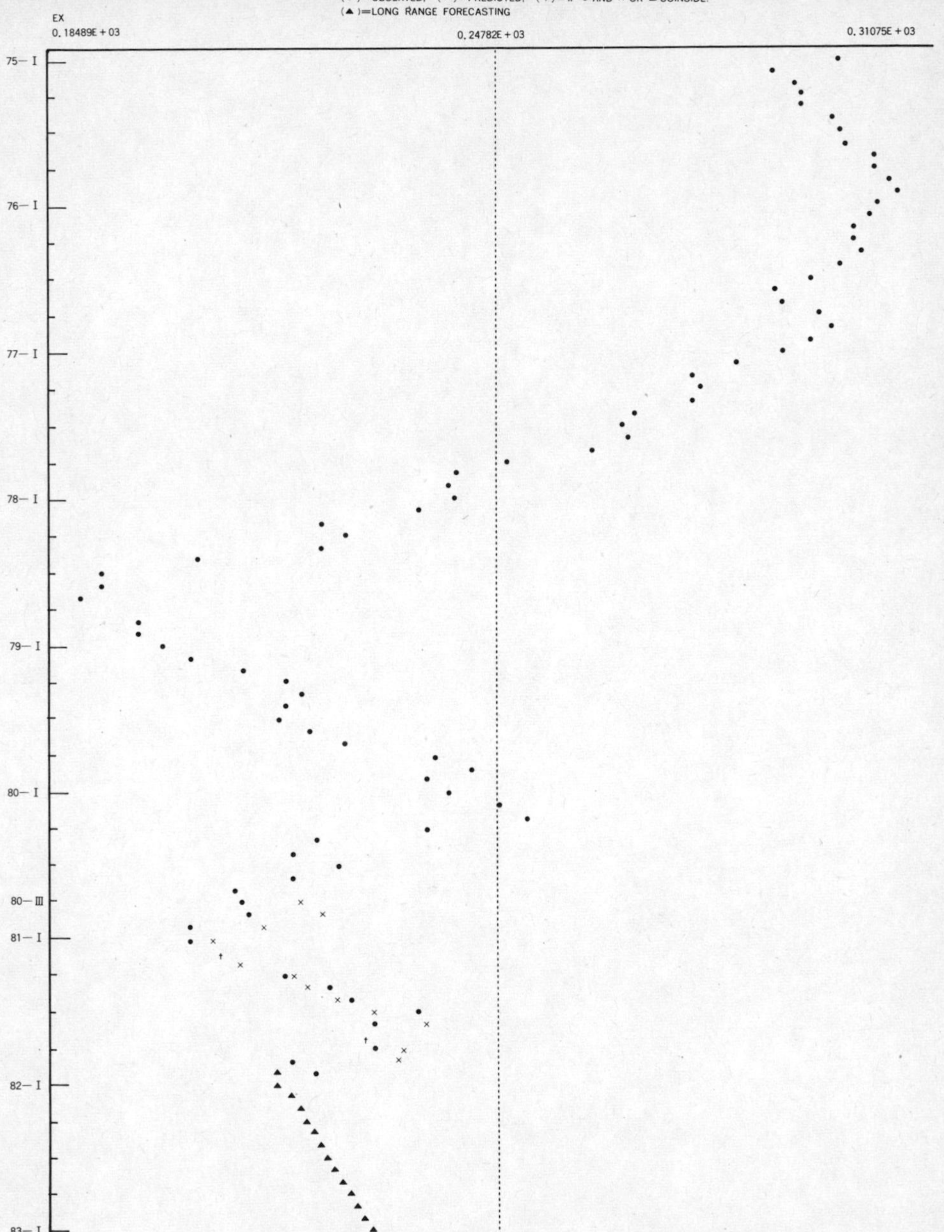
(●)=OBSERVED, (×)=PREDICTED, (†)= IF ● AND × OR ▲ COINSIDE.
(▲)=LONG RANGE FORECASTING
EX
0.18489E+03
0.24782E+03
0.31075E+03
75— I
76— I
77— I
78— I
79— I
80— I
80— III
81— I
82— I
83— I

Fig. 10

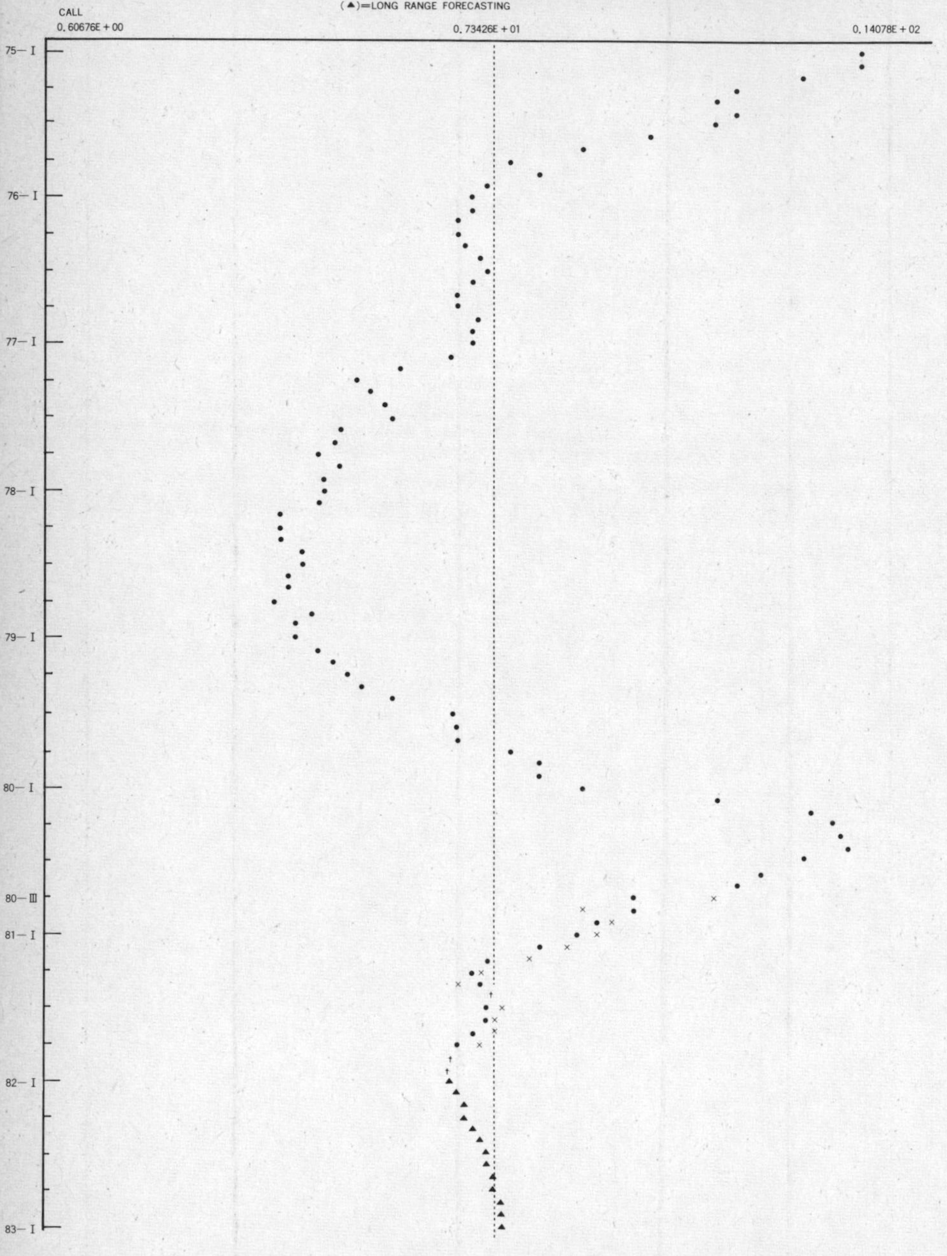
(●)=OBSERVED, (×)=PREDICTED, (†)=IF ● AND × OR ▲ COINSIDE.
(▲)=LONG RANGE FORECASTING
CALL
0.60676E+00
0.73426E+01
0.14078E+02
75—I
76—I
77—I
78—I
79—I
80—I
80—III
81—I
82—I
83—I

Fig. 11

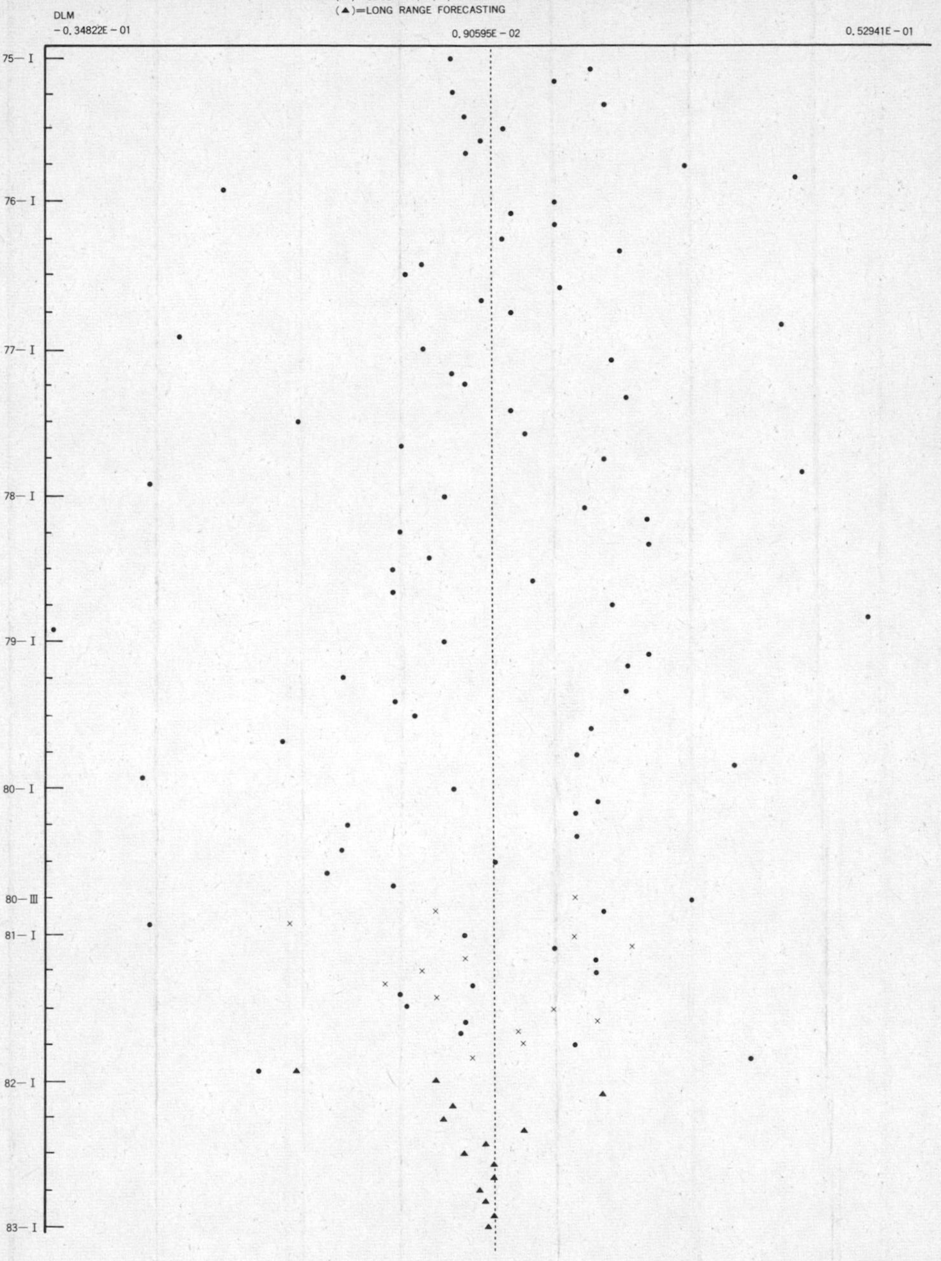
(●)=OBSERVED, (×)=PREDICTED, (†)=IF ● AND × OR ▲ COINSIDE.
(▲)=LONG RANGE FORECASTING
DLM
− 0. 34822E − 01
0. 90595E − 02
0. 52941E − 01
75− I
76− I
77− I
78− I
79− I
80− I
80− III
81− I
82− I
83− I

Fig. 12

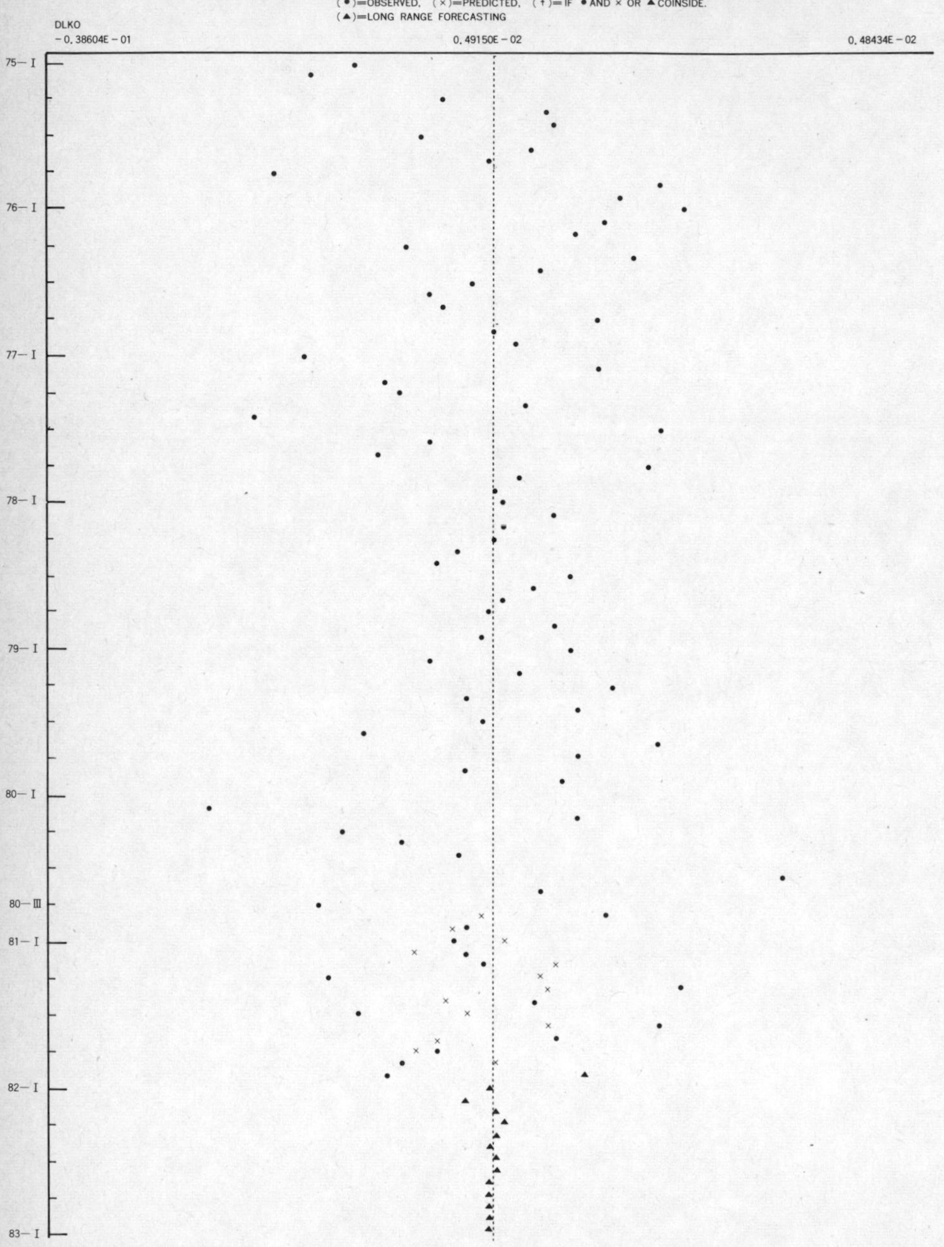
(●)=OBSERVED, (×)=PREDICTED, (†)=IF ● AND × OR ▲ COINSIDE.
(▲)=LONG RANGE FORECASTING
DLKO
-0.38604E-01
0.49150E-02
0.48434E-02
75-I
76-I
77-I
78-I
79-I
80-I
80-III
81-I
82-I
83-I

Fig. 13

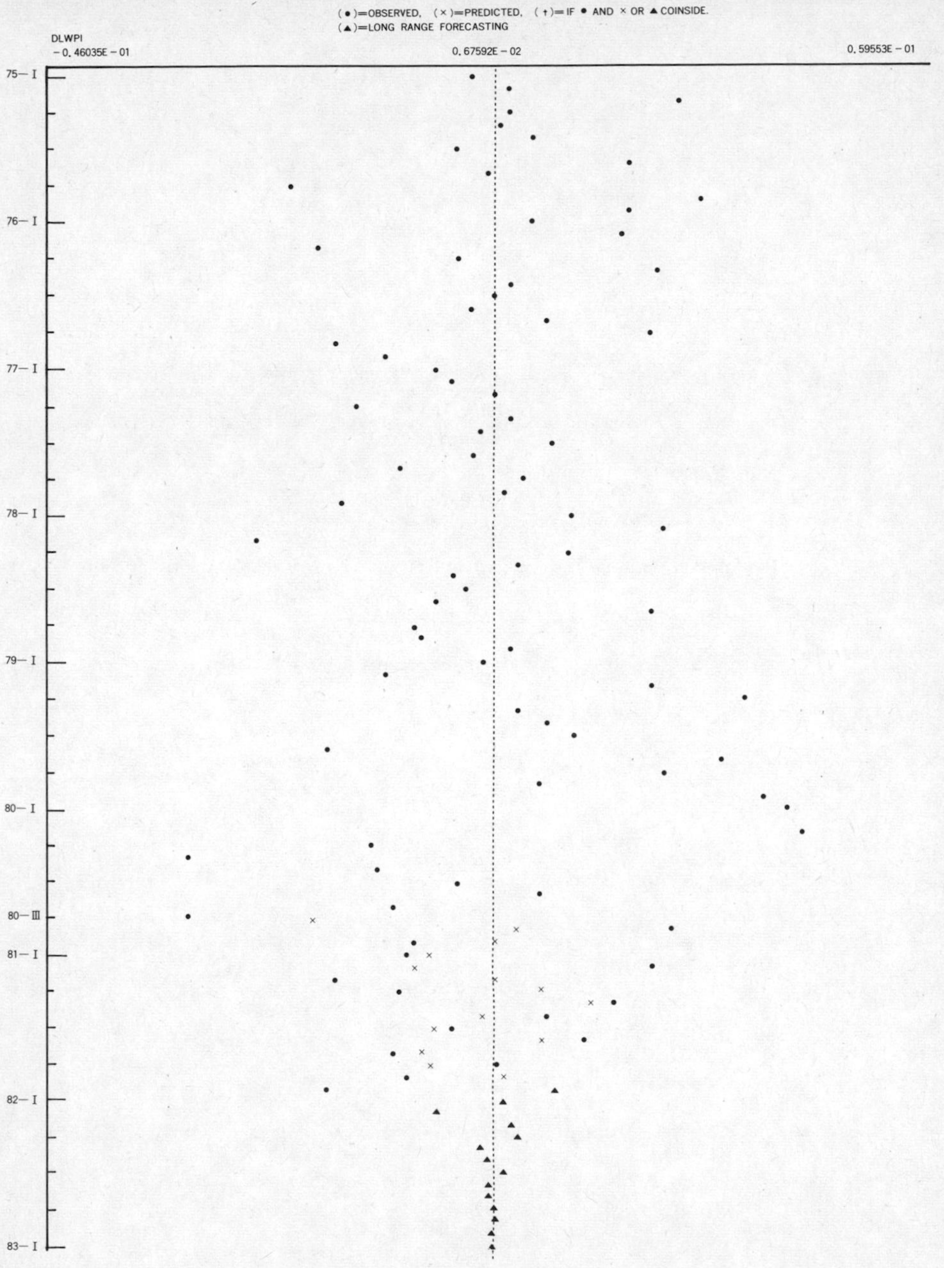
(●)=OBSERVED, (×)=PREDICTED, (+)=IF ● AND × OR ▲ COINSIDE.
(▲)=LONG RANGE FORECASTING
DLWPI
-0.46035E-01
0.67592E-02
0.59553E-01
75-I
76-I
77-I
78-I
79-I
80-I
80-III
81-I
82-I
83-I

MATHEMATICAL APPENDICES

A.1 Solutions of Difference Equations

A difference equation generates a solution or an output sequence, $\{y_n\}$, from another sequence $\{x_n\}$, called input or forcing terms. In general some regularity conditions need be imposed to ensure the uniqueness and stability of the solution sequences. Any sequence of exogenous variables can serve as an input sequence. Often, predetermined or lagged endogenous variables appear as parts of the forcing terms. A given input sequence $\{x_n\}$ is transformed into the solution sequence $\{y_n\}$ via the relationships embodied in the difference equation. We consider only linear difference equations where the transforming relations are linear: Suppose the system is initially at rest.* A constant c times $\{x_n\}$, i.e., the input sequence $\{cx_n\}$ then produces the solution sequence $\{cy_n\}$. If two input sequences $\{x_n\}$ and $\{\tilde{x}_n\}$ respectively generate $\{y_n\}$ and $\{\tilde{y}_n\}$ as the solution sequences, then the sequence $\{x_n+\tilde{x}_n\}$ produces $\{y_n+\tilde{y}_n\}$ as the corresponding solution sequence. This is known as the superposition principle for linear systems.

Solution of Linear Difference Equations

We usually write difference equations with the input functions on the right-hand side, all others on the left. The scalar difference equation $\alpha(L)y_t = x_t$, where x_t and y_t are real numbers and $\alpha(L) = 1 + \alpha_1 L + \ldots + \alpha_p L^p$, is one such example where x_t is the forcing term or input function at time t. The right-hand side may take a more complicated form such as $\beta(L)x_t$ for some polynominal $\beta(L)$, the point being that the right-hand side (RHS) depends on the known function x_t while y_t is to be solved for.

---

* No input sequence is present and only $\{0\}$ appears as the solution sequence, i.e., $x_n = 0$ and $y_n = 0$ for all $n \geq 0$.

Solutions of linear difference equations are made up of two parts; the homogeneous part, i.e., solutions when the right-hand side of the equation is zero and of the inhomogeneous part, i.e., solutions due to a non-zero input sequence. The former arises if the system is not at rest, and is called the zero-input solution. The latter is called the zero-state solution in the system literature. Several solution methods are available. We discuss a transform method and a direct i.e., a time-domain solution method.

1st Order Equation: $|\lambda| < 1$

An example will illustrate our procedure. Suppose that we have a first order difference equation, i.e., an equation with a single lag term

(1) $$(1-\lambda L)y_t = x_t.$$

Its homogeneous part is $(1-\lambda L)y_t = 0$. The expression $y_t = \lambda^t c$ is a solution where c is some constant. We can verify that it is a solution by substituting this expression back into the equation: $(1-\lambda L)\lambda^t c = \lambda^t c - \lambda\lambda^{t-1}c = 0$. To fix the constant, we need to specify the value of the solution at a point, usually at the initial time $t = 0$, i.e., y_0 is given as an initial condition. With this initial condition, the solution is unambiguously determined to be $y_t = \lambda^t y_0$. Alternatively, the solution may be fixed at another time such as a terminal time T. This terminal or boundary condition fixes the solution to be $y_t = \lambda^{(t-T)}y_T$. These two expressions are equivalent. Knowing y_T we can determine y_0 and conversely. Let $y_0 = 0$. The solution with the non-zero input is given by

(2) $$y_t = \sum_{\tau=1}^{t} \lambda^{t-\tau}x_\tau = \sum_{0}^{t-1} \lambda^s x_{t-s}.$$

To see that this satisfies the difference equation, rewrite y_t as $y_t = \Sigma_{\tau=1}^{t-1} \lambda^{t-\tau}x_\tau + x_t$, and note that the first term can be written as $\Sigma_{\tau=1}^{t-1} \lambda^{t-\tau}x_\tau = \lambda(\Sigma_{\tau=1}^{t-1} \lambda^{t-1-\tau}x_\tau)$ where the expression in parentheses is recognized as y_{t-1} from the supposed solution from (2) establishing that $y_t = x_t + \lambda y_{t-1}$.

The general solution is then made up of these two solutions

(3) $$y_t = \lambda^t y_0 + \sum_{\tau=1}^{t} \lambda^{t-\tau} x_\tau \quad \text{or} \quad \lambda^t y_0 + \sum_{s=0}^{t-1} \lambda^s x_{t-s}.$$

The first part is the zero-input solution and the second zero-state or zero-initial condition solution.

As t approaches infinity, the zero-input part of the solution remains bounded for $y_0 \neq 0$ if and only if $|\lambda| \leq 1$. With the bounded exogenous sequence, i.e., $|x_\tau| \leq M$ for all τ and for some M, the second term remains bounded as t goes to infinity if and only if $|\lambda| \leq 1$. Such a system is said to be bounded-input bounded-output (BIBO) stable. If $|\lambda| < 1$, then $y_t \to 0$ as $t \to \infty$ for all bounded input sequences. The system is then called asymptotically stable. With $|\lambda| > 1$, y_0 must be zero for y_t to remain bounded. Even when $|\lambda| > 1$, the second term can remain bounded if $|x_\tau|$ goes to zero sufficiently fast.

Formally, the solution can be obtained by writing the original equation as $y_t = \frac{1}{1-\lambda L} x_t$. Expand $(1-\lambda L)^{-1}$ as an infinite sum $\Sigma_{\tau=0}^{\infty} (\lambda L)^\tau$ assuming that this sum is finite. Then we can write y_t as

$$\begin{aligned} y_t &= \sum_{\tau=0}^{\infty} \lambda^\tau L^\tau x_t \\ &= \sum_{\tau=0}^{t-1} \lambda^\tau L^\tau x_t + \sum_{\tau=t}^{\infty} \lambda^\tau L^\tau x_t \\ &= \sum_{\tau=0}^{t-1} \lambda^\tau x_{t-\tau} + \lambda^t \sum_{\tau=t}^{\infty} \lambda^{\tau-t} x_{t-\tau} \\ &= \sum_{\tau=0}^{t-1} \lambda^\tau x_{t-\tau} + \lambda^t \{ \sum_{s=0}^{\infty} \lambda^s x_{-s} \}. \end{aligned}$$

If we identity the bracketed expression in the second term as y_0, i.e., $y_0 = \Sigma_{s=0}^{\infty} \lambda^s x_{-s}$, then we can write $y_t = \Sigma_{\tau=0}^{t-1} \lambda^\tau x_{t-\tau} + \lambda^t y_0$. This is exactly the solution we obtained earlier. The above manipulation is legitimate then if $\{x_t\}$ is a bounded sequence and if $|\lambda| < 1$. As a special case, suppose $x_t = a$ for all t. Then $(1-\lambda L)^{-1} a = (\Sigma_{i=0}^{\infty} \lambda^i) a = (1-\lambda)^{-1} a$ if $|\lambda| < 1$. The solution becomes $y_t = a/(1-\lambda)$ or $y_t = \lambda(\Sigma_{\tau=0}^{t-1} \lambda^{\tau})a + \lambda^t y_0 = (1-\lambda^t)a/(1-\lambda) + \lambda^t y_0$, where $y_0 = a/(1-\lambda)$.

The expansion into an infinite series implicitly assumes that the values going back to $-\infty$ are available. When the series starts up from a finite past point, the value there can be used as initial conditions.

Transform Methods

The procedure of the previous section for solving difference equations by formal expansion of $(1-\lambda L)^{-1}$ is related to the solution method by z-transforms. (The z-transform is the difference equation counterpart of the Laplace transform method of solving differential equations.) The z-transform of a sequence $\{x_n, n \geq 0\}$ is defined to be $X(z) = \Sigma_{n=0}^{\infty} x_n z^{-n}$. This is a formal series in which z serves as a place marker. By examining the coefficient of z^{-7} we can identify x_7, for example. A brief discussion of z-transforms and its relations to the lag and Fourier transforms are found in Appendix.

Denote the z-transform of the solution (sequence) by Y(z), i.e., $Y(z) = \Sigma_{n=0}^{\infty} y_n z^{-n}$. To obtain the z-transform of $\{Ly_n\}$, we define another sequence $\{h_n\}$ by $h_n = Ly_n = y_{n-1}$. Its z-transform is $H(z) = \Sigma_{n=0}^{\infty} h_n z^{-n} = \Sigma_{n=0}^{\infty} y_{n-1} z^{-n} = y_{-1} + \Sigma_{m=0}^{\infty} y_m z^{-m-1} = y_{-1} + z^{-1}Y(z)$. Consider the difference equation (1). We have just shown that the z-transform of the sequence $\{(1-\lambda L)y_t\} = \{y_t - \lambda y_{t-1}\}$ equals $Y(z) - \lambda\{y_{-1}+z^{-1}Y(z)\}$. This must equal the z-transform of the RHS of the difference equation, i.e., X(z). Equating the two, the resulting equation $Y(z) - \lambda y_{-1} - \lambda z^{-1}Y(z) = X(z)$ can be solved for Y(z) to yield

$$Y(z) = \lambda y_{-1}/(1-\lambda z^{-1}) + X(z)/(1-\lambda z^{-1}).$$

By examining the coefficients of the power z^{-t} on both sides we can obtain the expression for y_t. On the left-hand side (LHS), it is simply y_t by the construction of the z-transform. From the first term of the RHS we get $\lambda y_{-1}\lambda^t$

$= \lambda^{t+1} y_{-1}$. The second term of the RHS is $(x_0 + x_1 z^{-1} + x_2 z^{-2} + \ldots)(1 + \lambda z^{-1} + \lambda^2 z^{-2} + \ldots)$. Collecting the terms of the power z^{-t} yields $\lambda^t x_0 + \lambda^{t-1} x_1 + \ldots + x_t$. Hence we recover the solution we earlier obtained by this method as well:

$$y_t = \lambda^{t+1} y_{-1} + \sum_{\tau=0}^{t} \lambda^\tau x_{t-\tau}$$

$$= \lambda^t y_0 + \sum_{0}^{t-1} \lambda^\tau x_{t-\tau},$$

where $y_0 = \lambda y_{-1} + x_0$ has been substituted out. As in the earlier method, if we wish to have well-defined z-transforms such as $X(z)$ or $Y(z)$, we can regard z as the complex variable and suitably restrict the domain of z over which the series are convergent.

1st Order Equation: $|\lambda| > 1$

Ordinarily, we solve difference equations forward in time from some initial time. In economics, however, we often want to solve difference equations backward in time relating the solution values to a future time instant such as the end of a planning horizon as we do in dynamic programing. Our earlier method treated time as flowing forward, i.e., knowing y_0 we determined y_t using values of input sequences $x_1, x_2, \ldots, x_t$, $t > 0$. When we specify a value of the solution at some future time $T > 0$, we are solving the difference equation backward in time to obtain y_t, $t < T$, from the specified boundary or terminal condition y_T.

The zero-input solution of $(1-\lambda L) y_t = x_t$ is

$$y_t = \lambda^{t-T} y_T = (1/\lambda)^{T-t} y_T. \tag{4}$$

We see that y_t goes to zero as $T \to \infty$ if $|\lambda| > 1$. If $|\lambda| < 1$, then y_t diverges as T approaches infinity. To obtain the zero-state or zero-initial (now zero-terminal) condition part of the solution, we measure time backward from T by changing the time variable from t to $s = T - t$, and rename the variables:

$h_s = y_{T-s}$, $u_s = x_{T-s}$. The difference equation in these new variables is $h_s - \lambda h_{s+1} = u_s$ or $h_{s+1} - (1/\lambda)h_s = -(1/\lambda)u_s$. This is the type of equations we discussed in the previous section because $1/|\lambda| < 1$. We can write its zero-input solution as $h_s = (1/\lambda)^s h_0$. In the original variables we recover $y_{T-s} = (1/\lambda)^s y_T$ or $y_t = (1/\lambda)^{T-t} y_T$. Its zero-state solution is $h_s = -(1/\lambda)\Sigma_{i=0}^{s-1}(1/\lambda)^i \cdot u_{s-i-1}$.* Converting back to the original variables, this expression becomes $y_{T-s} = -(1/\lambda)\Sigma_{i=0}^{s-1}(1/\lambda)^i x_{T+1-s+i}$. Renaming the time variable T-s by t, we rewrite this solution as

(5)
$$y_t = -\frac{1}{\lambda}\sum_{n=0}^{T-1-t}(-\frac{1}{\lambda})^n x_{t+n+1}.$$

It shows that y_t is affected by the current and future inputs, $x_{t+1}, \ldots, x_T$ rather than by current and past inputs as in the previous section.

The general solution combines (4) and (5):

(6)
$$y_t = (\frac{1}{\lambda})^{T-t} y_T - \frac{1}{\lambda}\sum_{i=0}^{T-1-t}(-\frac{1}{\lambda})^i x_{t+1+i}.$$

(Of course (6) can be directly obtained by iterating backward from t = T.) This form of the solution relates the current value of the solution y_t to the terminal value y_T and the exogenous (input) variables that will occur between now, t, and the future, T-1. Comparing (2) with (6), we note that $|1/\lambda|$ rather than $|\lambda|$ must be less than 1 if y_t remains bounded as $T \to \infty$. By letting T approach infinity (5) becomes

(7)
$$y_t = -(\frac{1}{\lambda})\sum_{s=t}^{\infty}(\frac{1}{\lambda})^{s-t} x_{s+1} = -(\frac{1}{\lambda})\sum_{i=0}^{\infty}(\frac{1}{\lambda})^i x_{t+1+i}.$$

Formally, this form of solution can be obtained from the original difference equation without these changes of variables by expanding $(1-\lambda L)^{-1}$ not as the power series in (λL) by rather as a formal power series in $(\lambda L)^{-1}$: $(1-\lambda L)^{-1}$

---

* Compared with (1), the time index of u is off by one because $(1-\lambda L)h_{s+1} = -(1/\lambda)u_s$ rather than $(1-\lambda L)h_s = -(1/\lambda)u_s$.

$= -(\lambda L)^{-1}\{1-(\lambda L)^{-1}\}^{-1}$ which becomes $-\Sigma_{i=1}^{\infty}(\lambda L)^{-i}$. Then the solution of $(1-\lambda L)y_t = x_t$ is written as

$$
\begin{aligned}
(7') \qquad y_t &= (1-\lambda L)^{-1}x_t \\
&= -\sum_{i=1}^{\infty}(\lambda L)^{-i}x_t \\
&= -\lambda^{-1}\sum_{i=1}^{\infty}\lambda^{-i}x_{t+i+1}
\end{aligned}
$$

because L^{-1} now stands for a forward shift of time index; $L^{-1}x_t = x_{t+1}$. This expression is exactly (7).

By breaking up the infinite sum as

$$
(8) \qquad y_t = -(\frac{1}{\lambda})\sum_{i=0}^{T-1-t}(\frac{1}{\lambda})^i x_{t+1+i} - (\frac{1}{\lambda})\sum_{i=T-t}^{\infty}(\frac{1}{\lambda})^i x_{t+1+i}
$$

and rearrange the second term as

$$
\begin{aligned}
-(\frac{1}{\lambda})\sum_{i=T-t}^{\infty}(\frac{1}{\lambda})^i x_{t+1+i} &= -(\frac{1}{\lambda})\sum_{j=0}^{\infty}(\frac{1}{\lambda})^{T-t+j}x_{T+1+j} \\
&= (\frac{1}{\lambda})^{T-t}y_T
\end{aligned}
$$

we see that (7) can be put as (6), provided the infinite sum is absolutely convergent (for example, x's being bounded and $|\lambda| > 1$). The z-transform method of solving (1) for $|\lambda| > 1$ involves the same sort of manipulations.

2nd Order Equation

We can solve higher-order difference equations for scalar variables in several ways. The most systematic and theoretically satisfying way is to convert them into first-order difference equations for suitably constructed vectors. These vectors are the state vectors (of the system governed by the difference equations in question). We can then appeal to a body of linear system theory to obtain insight into solution behavior. Because this tack requires some knowledge of system theory (as summarized in Aoki [1976; Part I], for example), which is probably not familiar to the economic profession, we first proceed as follows, using z-transform or Lag-transform to tackle a second-order difference equation. We wish to solve

(9) $$a(L)y_t = x_t$$

where

$$a(L) = 1 + a_1 L + a_2 L^2$$

$$= (1-\lambda_1 L)(1-\lambda_2 L).$$

Formally (9) leads to the solution

$$y_t = \frac{1}{(1-\lambda_1 L)(1-\lambda_2 L)} x_t.$$

Expanding $\frac{1}{(1-\lambda_1 L)(1-\lambda_2 L)}$ into partial fraction, $\frac{1}{1-\lambda_2/\lambda_1} \frac{1}{1-\lambda_1 L} + \frac{1}{1-\lambda_1/\lambda_2} \frac{1}{1-\lambda_2 L}$, y_t can be expressed as

$$y_t = \frac{1}{1-\lambda_2/\lambda_1} \frac{x_t}{1-\lambda_1 L} + \frac{1}{1-\lambda_1/\lambda_2} \frac{x_t}{1-\lambda_2 L}.$$

Now if $|\lambda_1|$, $|\lambda_2|$ are both less than one, then our solution of the first order equation (2) immediately leads to the solution of (9)

$$y_t = \frac{1}{1-\lambda_2/\lambda_1} \sum_{i=0}^{\infty} \lambda_1^i x_{t-i} + \frac{1}{1-\lambda_1/\lambda_2} \sum_{i=0}^{\infty} \lambda_2^i x_{t-i}.$$

Now, by identifying y_{01} with $\Sigma_{i=1}^{\infty}(1-\lambda_2/\lambda_1)^{-1}\lambda_1^i x_{-i}$ and y_{02} with $\Sigma_{i=1}^{\infty}(1-\lambda_1/\lambda_2)^{-1}\cdot \lambda_2^i x_{-i}$ we can write the above equivalently as

(10) $$y_t = \sum_{i=0}^{t} h_i x_{t-i} + \lambda_1^t y_{01} + \lambda_2^t y_{02}$$

where

$$h_i = (1-\lambda_2/\lambda_1)^{-1}\lambda_1^i + (1-\lambda_1/\lambda_2)^{-1}\lambda_2^i.$$

This expression corresponds to (3). We note that we must now have two conditions to fix the two constants (initial conditions) y_{01} and y_{02}. They are written here as two components of an initial condition vector.

In (10), the zero-input solutions are represented by the last two terms.

Because $(1-\lambda_1 L)\lambda_1^t = 0$ and $(1-\lambda_2 L)\lambda_2^t = 0$, we have no doubt that $(1-\lambda_1 L)(1-\lambda_2 L)\cdot(c_1\lambda_1^t+c_2\lambda_2^t) = 0$ for any c_1 and c_2. Zero-input solutions hence general solutions of second-order equations need two conditions to fix the solution uniquely. A specific solution is picked by fixing the solution sequence at any two points; one condition could be specified at t = 0 and the other at t = T (some future or terminal time).

The sequence $\{h_i\}$ is the impulse response sequence which represents the dynamic (i > 0) and impact (i = 0) multiplier effects of the exogenous variable x on y. If both $|\lambda_1|$ and $|\lambda_2|$ are greater than one, then our solution (6) or (7) suggest that we write 1/a(L) as

$$\frac{1}{(1-\lambda_1 L)(1-\lambda_2 L)} = -\left\{\frac{1}{1-\lambda_2/\lambda_1}\frac{(\lambda_1 L)^{-1}}{1-(\lambda_1 L)^{-1}} + \frac{1}{1-\lambda_1/\lambda_2}\frac{(\lambda_2 L)^{-1}}{1-(\lambda_2 L)^{-1}}\right\}$$

so that y_t is formally written as

$$y_t = -(\lambda_1-\lambda_2)^{-1}\sum_{i=0}^{\infty}\left(\frac{1}{\lambda_1}\right)^i x_{t+1+i} - (\lambda_2-\lambda_1)^{-1}\sum_{i=0}^{\infty}\left(\frac{1}{\lambda_2}\right)^i x_{t+1+i}.$$

Or if we wish, we can break up each of the two infinite sums into two parts as in (8) and write the above as

$$y_t = \left(\frac{1}{\lambda_1}\right)^{T-t} y_{T1} + \left(\frac{1}{\lambda_2}\right)^{T-t} y_{T2} - (\lambda_1-\lambda_2)^{-1}\sum_{i=0}^{T-1-t}\left(\frac{1}{\lambda_1}\right)^i x_{t+1+i}$$
$$-(\lambda_2-\lambda_1)^{-1}\sum_{i=0}^{\infty}\left(\frac{1}{\lambda_2}\right)^i x_{t+1+i}.$$

Again two constants need be specified. They are expressed here as two components of a vector specified at the terminal time T.

Suppose $|\lambda_1| < 1 < |\lambda_2|$. Then it is sometimes convenient to express y_t as

$$(1-\lambda_1 L)y_t = x_t/(1-\lambda_2 L)$$
$$= -\lambda_2^{-1}\sum_{i=0}^{T-1-t}\lambda_2^{-i}x_{t+1+i} + c\left(\frac{1}{\lambda_2}\right)^{T-t}$$

where c is to be determined by a condition specified at T. Or letting $T \to \infty$, we have

$$(1-\lambda_1 L)y_t = -\lambda_2^{-1} \sum_{i=0}^{\infty} \lambda_2^{-i} x_{t+1+i}.$$

Then y_t is expressible as

$$y_t = \lambda_1 y_{t-1} - \lambda_2^{-1} \sum_{i=0}^{\infty} \lambda_2^{-i} x_{t+1+i}.$$

This form is useful if y_{t-1} is known. Then y_t is determined by the future stream of x's. When x is stochastic, this form or slightly altered version of it often appears as the one-step ahead prediction formula of (now random) y_t, given y_{t-1}. Such examples are found in Sargent [1979] and elsewhere in this lecture notes as well.

State Space Representation

We now solve (9) using state space. Define a vector s_t by $s_t = (y_t, y_{t-1})'$. The second-order difference equation (9) then is equal to

$$\begin{pmatrix} y_t \\ y_{t-1} \end{pmatrix} = \begin{pmatrix} -a_1 & -a_2 \\ 1 & 0 \end{pmatrix} \begin{pmatrix} y_{t-1} \\ y_{t-2} \end{pmatrix} + \begin{pmatrix} 1 \\ 0 \end{pmatrix} x_t$$

or

$$s_t = As_{t-1} + bx_t \tag{11}$$

where

$$A = \begin{pmatrix} -a_1 & -a_2 \\ 1 & 0 \end{pmatrix} \quad \text{and} \quad b = \begin{pmatrix} 1 \\ 0 \end{pmatrix}.$$

The characteristic polynomial of A, $|\lambda I - A|$, is a second-order polynomial in λ, $\lambda^2 + a_1\lambda + a_2$. Its roots are the eigenvalues of A which are exactly given by λ_1 and λ_2 we have earlier used to factor the lag polynomial a(L) of (9). Now (11) produces a vector version of the 1st order difference equation in lag-operator form

(12) $$(I-AL)s_t = bx_t.$$

Here AL is understood to equal $\begin{pmatrix} -a_1L & -a_2L \\ L & 0 \end{pmatrix}$. This corresponds to (1) we have earlier discussed. Eq (12) is formally solved as $s_t = (I-AL)^{-1}bx_t$. From the matrix identity $(I-AL)^{-1} = \text{adj}(I-AL)/|I-AL|$ where $|I-AL| = 1+a_1L+a_2L^2 = (1-\lambda_1L)\cdot(1-\lambda_2L)$, we can write s_t as

$$s_t = \frac{1}{(1-\lambda_1L)(1-\lambda_2L)}\begin{pmatrix} 1 \\ L \end{pmatrix}x_t.$$

Componentwise, this is nothing but $y_t = [(1-\lambda_1L)(1-\lambda_2L)]^{-1}x_t$ and $y_{t-1} = [(1-\lambda_1L)\cdot(1-\lambda_2L)]^{-1}Lx_t$.

The solution of (1) can be put then as

(13) $$s_t = A^ts_0 + \sum_{\tau=1}^{t} A^{t-\tau}bx_\tau$$

$$= A^ts_0 + \sum_{\tau=0}^{t-1} A^\tau bx_{t-\tau}.$$

Suppose that A has two linearly independent eigenvectors u_1 and u_2 so that $A[u_1, u_2] = [u_1, u_2]\Lambda$ where $\Lambda = \text{diag}(\lambda_1, \lambda_2)$.

Define v_1 and v_2 by $[u_1, u_2]^{-1} = \begin{pmatrix} v_1' \\ v_2' \end{pmatrix}$. The matrix A is then expressible as

$$A = [u_1, u_2]\begin{pmatrix} v_1' \\ v_2' \end{pmatrix} = \sum_{i=1}^{2} \lambda_iu_iv_i'.$$

The vector v_i' is the left row eigenvector, $v_i'A = \lambda_iv_i'$. This is an example of the spectral decomposition representation discussed in Appendix. We note that $v_i'u_j = \delta_{ij}$ from the construction. Because of this, the power A^n has $\Sigma_{i=1}^2 \lambda_i^nu_iv_i'$ as its spectral representation. Then (12) can be written as

$$s_t = \sum_{i=1}^{2} \lambda_i^tu_i(v_i's_0) + \sum_{\tau=0}^{t-1}\sum_{i=1}^{2} \lambda_i^\tau u_i(v_i'b)x_{t-\tau}.$$

We note that λ_2^t is absent from s_t if $v_2's_0$ is zero. If $v_2's_0$ is zero, then λ_1^t is absent from the first summation. We see that the initial condition vector s_0 which is orthogonal to v_i does not excite the i-th term. This observation generalizes to n-dimensional problems. We refer to the i-th term as the i-th mode of the dynamic system.

A.2 Geometry of Weakly Stationary Stochastic Sequences

Economic agents often face a noisy environment and must extract information useful for his decision problems from it. Totality of mean-zero random vectors with finite variances can be made a Hilbert space by defining an inner product of two members x an y by $(x, y) = Ex'y$. Two random vectors are orthogonal if the inner product is zero, i.e., if they are uncorrelated.

Suppose y_t is observed which is related to a basic or elementary random sequence, ε_t, $E\varepsilon_t = 0$, $E\varepsilon_t\varepsilon_s = \sigma^2\delta_{t,s}$, by a moving average process

(1) $$y_t = \phi(L)\varepsilon_t$$

where

$$\phi(L) = \sum_{j=0}^{\infty} \phi_j L^j, \quad \phi_0 = 1.$$

For this to generate a covariance stationary process, we assume that $\Sigma_{j=0}^{\infty} \phi_j^2$ is finite.

Let us consider predicting y_{t+m}, $m > 0$ based on information set $\tilde{I}_t = \{y_t, y_{t-1}, y_{t-2}, \dots \}$, as a typical or one of the basic problems of information extraction faced by economic agents. We consider only a linear prediction formula in which y_{t+m} is estimated by a linear combination of current and past y's.

Here we assume that $\phi(L)$ is invertible in the sense that $\phi(L)^{-1} = \psi(L)$ exists such that $\psi(L) = \Sigma_{j=0}^{\infty} \psi_j L^j$, $\Sigma_{j=0}^{\infty} \psi_j^2 < \infty$. In other words, the $\{y_t\}$ process of (1) can equivalently be expressed as an autoregressive process

(2) $$\psi(L)y_t = \varepsilon_t.$$

A simple example $y_t = \varepsilon_t - \varepsilon_{t-1}$ shows that not all moving average processes are invertible. However, invertible processes do constitute an important class of stochastic processes. For such processes, the information set $\tilde{I}_t$ is equivalent to the one containing $I_t = \{\varepsilon_t, \varepsilon_{t-1}, \dots \}$. Then a linear prediction of y_{t+m} based on I_t is of the form

$$y_{t+m|t} = A(L)\varepsilon_t$$

where

$$A(L) = a_0 + a_1 L + a_2 L^2 + \ldots .$$

The best such prediction is, by definition, the one that minimizes the prediction error variance

$$\sigma_y^2 = E[(y_{t+m} - \hat{y}_{t+m|t})^2 | I_t]$$

$$= E[\{\phi(L)\varepsilon_{t+m} - A(L)\varepsilon_t\}^2 | I_t].$$

Separate out the future ε's from those in the information set I_t by writing

$$\phi(L)\varepsilon_{t+m} = \sum_{j=0}^{m-1} \phi_j \varepsilon_{t+m-j} + \sum_{j=m}^{\infty} \phi_j \varepsilon_{t+m-j}$$

$$= \sum_{j=0}^{m-1} \phi_j \varepsilon_{t+m-j} + \phi_m(L)\varepsilon_t,$$

where

$$\phi_m(L) = \sum_{i=0}^{\infty} \phi_{m+i} L^i .$$

Then

$$\sigma_y^2 = \sum_{j=0}^{m-1} \phi_j^2 \sigma^2 + [\{\phi_m(L) - A(L)\}\varepsilon_t]^2 .$$

The choice $A(L) = \phi_m(L)$ clearly minimizes the σ_y^2, i.e.,

(3) $$y_{t+m|t} = \phi_m(L)\varepsilon_t$$

is the best least-squares predictor of y_{t+m} based on information contained in I_t.

Sometimes a notation or an (annihilator) operator $[\cdot]_+$ is used to express $\phi_m(L)$ as $[\phi(L)/L^m]_+$ where $[\cdot]_+$ collects only non-negative powers of L dropping all expressions with negative powers of L. Then we can write the best l.s. predictor as $[\phi(L)/L^m]_+\varepsilon_t$. Non-negative powers of L refer to current and past values. Recalling that z^{-1} corresponds to L, we can equivalently define $[\ ]_+$ as non-negative powers in z^{-1} for any expression expressed as (formal) power series in z^{-1}. For example, we write

$$F(z) = \sum_{-\infty}^{\infty} f_n z^{-n} = [F(z)]_+ + [F(z)]_-$$

where

$$[F(z)]_+ = \sum_{n=0}^{\infty} f_n z^{-n},$$

and

$$[F(z)]_- = \sum_{n=1}^{\infty} f_{-n} z^{n}.$$

The operation $[\ ]_+$ picks out causal (i.e., realizable) portion of the transfer function $[\ ]$. When basic or fundamental noise sequences are involved, $[\ ]_+$ realizes the orthogonal projection onto the subspace spanned by the data. Examine further the expression in (3). The variable to be predicted y_{t+m} can be written as $\phi(L)L^{-m}\varepsilon_t$ because $L^{-m}\varepsilon_t$ is ε_{t+m}. The best predictor, $[\phi(L)/L^m]_+\varepsilon_t$, drops from $\phi(L)L^{-m}\varepsilon_t$ the random variables $\varepsilon_{t+m}, \varepsilon_{t+m-1}, \ldots, \varepsilon_{t+1}$. Because $\varepsilon_{t+1}, \ldots, \varepsilon_{t+m}$ are uncorrelated with ε's in the information set I_t, the dropping of these uncorrelated random variables is equivalent to taking orthogonal projection of them onto the subspace spanned by ε's in I_t. The operator $[\phi(L)/L^m]_+$ is the orthogonal projection operator. Alternately put, the best predictor $y_{t+m|t}$ is such that the predictor error $y_{t+m} - y_{t+m|t}$ is orthogonal to (uncorrelated with) all the ε's in I_t, or by the equivalence of the subspace spanned by $\varepsilon_t, \varepsilon_{t-1}, \ldots$ with that spanned by $y_t, y_{t-1}, \ldots, y_{t+m}-y_{t+m|t}$ is orthogonal to $y_{t-\tau}$, $\tau = 0, 1, \ldots$.

This fact is sometimes referred to as the orthogonality principle: Consider a collection of random variables $x_1, \ldots, x_n$ which are used to estimate another random variable y in the mimimum mean square sense, i.e.,

$$\underset{\{a_i\}}{\text{Min}}\ (y-\Sigma a_i x_i,\ y-\Sigma a_i x_i) = \underset{\{a_i\}}{\text{Min}}\ E(y-\sum_{i=1}^{n} a_i x_i)^2.$$

Examining a slight change in a_j from its optimal value a_j^0, we note that the coefficients are optimal if and only if

$$0 = (y - \sum_{i=1}^{n} a_i^0 x_i,\ x_j), \quad j = 1, \ldots, n.$$

In words, the optimal estimation error, $y - \Sigma_{i=1}^{n} a_i^0 x_i$, is orthogonal to every vector $x_1, \ldots, x_n$. See Achieser [1956, Chapter 1], for example, for further discussion.

A.3 Principal Components

Definition The principal components of a p-dimensional random vector x with mean zero and covariance matrix X are defined as linear combinations of the components of the vector x

$$v = \Gamma' x$$

where Γ is the matrix made up of the (normalized) eigenvectors of X, $\Gamma = [\gamma_1, \gamma_2, \ldots, \gamma_p]$, i.e.,

$$X\Gamma = \Gamma\Lambda, \quad \Gamma'\Gamma = I$$

where $\Lambda = \text{diag}\ (\lambda_1, \ldots, \lambda_p)$.

The i-th component of the vector v, v_i, is called the i-th principal component of x.

This definition makes clear that the principal components are the coordinates of x with respect to the basis composed of the eigenvectors of Σ, because $x = \Gamma\xi$ implies that $v = \Gamma' x = \Gamma'\Gamma\xi = \xi$ if $\Gamma'\Gamma = I$, i.e., v is the representation of x with respect to the basis Γ.

The principal components are orthogonal because

$$\text{cov}\ (v) = \Gamma'\overline{xx'}\Gamma = \Gamma' X \Gamma = X.$$

The variance of v_i is the i-th eigenvalue of Σ. From $\text{tr}\ \Sigma = \Sigma_1^p\ \lambda_i$, the sum of all the variances remains fixed.

Optimality Properties

Principal components possess several optimal properties. For example, Rao [1964] discuss them. To illustrate, consider

$$\text{Max}\ (\text{var}\ h'x\colon\ h'h = 1).$$

Because $\text{var}\ h'x = h'\Sigma h$, the choice of h to be the normalized eigenvector of Σ with the largest eigenvalue λ_1 achieves this maximum. If the sum of the remaining eigenvalues $\gamma_2 + \ldots + \gamma_p$ is negligible compared with γ_1, then most of the variance of x is explained by $\gamma_1' x$, i.e., by the first principal component of x.

If not, we can continue the search for a small number of variables that account for most variance of x by choosing h'x to satisfy

$$\max (\text{var } h'x: \ h'h = 1, \ h'\Sigma\gamma_1 = 0).$$

The condition $h'\Sigma\gamma_1 = 0$ means that h'x is uncorrelated with $\gamma_1'x$. Because γ_2 is orthogonal to γ_1, $\gamma_2'x$ achieves the maximum.

Alternatively, the spectral decomposition of Σ

$$\Sigma = \sum_{i=1}^{p} \lambda_i \gamma_i \gamma_i'$$

shows that $\lambda_1\gamma_1\gamma_1'$ is the contribution by the first principal component and so on in explaining the total variance to Σ.

Another optimization problem that is solved by the principal components is to approximate Σ by another p × p matrix B of rank < p.

$$\min_B \|\Sigma - B\|$$

where $\|\Sigma - B\|^2 = \text{tr } (\Sigma - B)(\Sigma - B')'$, rank B = q < p.

Noting that Γ introduced earlier is orthogonal

$$\begin{aligned} \text{tr } (\Sigma - B)(\Sigma - B)' &= \text{tr } \Gamma\Gamma'(\Sigma - B)\Gamma\Gamma'(\Sigma - B)' \\ &= \text{tr } \Gamma'(\Sigma - B)\Gamma\Gamma'(\Sigma - B)' \\ &= \text{tr } (D - G)(D - G)' \end{aligned}$$

where $G = \Gamma'B\Gamma$ where rank G = rank B because Γ is nonsingular.

Then from the relation

$$\|\Sigma - B\|^2 = \sum_{i=1}^{p} (\lambda_i - g_{ii})^2 + \sum\sum_{i \neq j} g_{ij}^2,$$

the above expression is clearly minimized by choosing

$$g_{ii} = \lambda_i, \ i = 1, \ldots, q,$$

$$g_{ij} = 0, \ i \neq j,$$

i.e., $B = \Sigma_1^q \lambda_i\gamma_i\gamma_i'$.

We next show that the internally balanced model of Moore [1976] is an appli-

of the principal components.

Let

$$\mathfrak{C} = [B, AB, \ldots],$$

and

$$\underline{u}^t = \begin{bmatrix} \varepsilon_t \\ \varepsilon_{t-1} \\ \vdots \end{bmatrix}.$$

Then the zero-state solution of a linear dynamics excited by a sequence of random impulses is

$$y_t = \mathfrak{C}\underline{u}^t$$

where we choose

$$E\underline{u}^t\underline{u}^{t'} = I.$$

The output covariance matrix is

$$Ey_t y_t' = \mathfrak{C}E(\underline{u}^t\underline{u}^{t'})\mathfrak{C}'$$

$$= \mathfrak{C}\mathfrak{C}' = G_c\colon n\times n,$$

where G_c is the controllability grammian.

Let Γ be the matrix made up of n normalized eigenvectors of G_c

$$G_c\Gamma = \Gamma\Sigma$$

where $\Sigma = \text{diag}\ (\sigma_1^2, \ldots, \sigma_n^2)$.

The vector $\Gamma'y_t$ is the vector of the principal components,

$$v = \Gamma'y_t,$$

or

$$y_t = \Gamma v = \sum_1^n \gamma_i u_i.$$

For example, the first component of v is expressible as

$$v_1 = \gamma_1\mathfrak{C}\underline{u}^t,$$

and

$$\text{cov}\ v_1 = \gamma_1'\mathfrak{C}\mathfrak{C}'\gamma_1$$

$$= \sigma_1^2.$$

A.4 Fourier Transforms

Preliminary Notions

A periodic function of time indefinitely repeats a basic pattern defined over a finite interval such as [0, T] or [-T/2, T/2], i.e., $x(t+T) = x(t)$ where T is called the period.

Any periodic function has a Fourier series expansion

$$(1) \qquad x(t) = \sum_{n=-\infty}^{\infty} c_n e^{j2\pi nt/T}$$

provided the integral exists for the Fourier coefficient

$$(2) \qquad c_n = \frac{1}{T}\int_{-T/2}^{T/2} x(t)e^{-j2\pi nt/T}dt$$

and $\Sigma|c_n|^2 < \infty$.

Here the basic pattern is taken to be given over the interval [-T/2, T/2]. If the pattern is defined over [0, T] then, integrate over [0, T] to define the expansion coefficient.

More generally, the Fourier Transform of any function x(t) is defined by

$$(3) \qquad X(\omega) = \int_{-\infty}^{\infty} x(t)e^{-j\omega t}dt$$

if the integral exists, for example if $\int_{-\infty}^{\infty} |x(t)|^p dt$ is finite for $1 \leq p \leq 2$.

The original function is recoverable from the Fourier transform by

$$(4) \qquad x(t) = \frac{1}{2\pi}\int_{-\infty}^{\infty} X(\omega)e^{j\omega t}d\omega.$$

This is exact if $x(\cdot)$ is continuous. The right-hand side defines $\{x(t+0) + x(t-0)\}/2$ if $x(\cdot)$ is discontinuous at t. Comparing (1) with (4), $X(\omega)$ is seen to correspond to the Fourier series expansion coefficient. This correspondence can be made plausible by the following arguments: Use the segment of x(t) over [-T/2, T/2] to construct a periodic function $x_T(t)$. Its Fourier series expansion is as in (1), i.e.,

$$x_T(t) = \sum_{-\infty}^{\infty} c_n(T)e^{j2\pi nt/T}, \tag{5}$$

where

$$c_n(T) = \frac{1}{T}\int_{-T/2}^{T/2} x(t)e^{-j2\pi nt/T}dt.$$

Let $X_T(2\pi n/T) = Tc_n(T)$, and denote $2\pi/T$ by $\Delta\omega$. Using these symbols, (5) is written as

$$x_T(t) = \frac{\Delta\omega}{2\pi}\sum_{n=-\infty}^{\infty} X_T(2\pi n/T)e^{j2\pi nt/T}, \tag{5'}$$

which has a form of series approximation to an integral. Letting $T \to \infty$ and assuming the convergence of the infinite sum

$$\Delta\omega \sum_{-\infty}^{\infty} X_T(2\pi n/T)e^{j\frac{2\pi n}{T}t} \int_{-\infty}^{\infty} X(\omega)e^{j\omega t}d\omega,$$

we see that as T approaches infinity

$$x_T(t) \to x(t) = \frac{1}{2\pi}\int_{-\infty}^{\infty} X(\omega)e^{j\omega t}d\omega.$$

Time Series Data

So far we have treated $x(\cdot)$ as a continuous function of time. Now suppose a time series is given, $\{x_n, n = 0, \pm 1, \pm 2, \ldots\}$. If we think of x_n as $x(nT)$ of some function $x(\cdot)$ with a sampling period T, then (4) evaluated as $t = nT$ gives us

$$x_n = \frac{1}{2\pi}\int_{-\infty}^{\infty} X(\omega)e^{jn\omega T}d\omega.$$

Breaking up the interval of integration into segments of length $2\pi/T$ each, write the above as

$$x_n = \frac{T}{2\pi}\int_{-\infty}^{\infty} \frac{1}{T}\int_{(2m-1)\pi/T}^{(2m+1)\pi/T} X(\omega)e^{jn\omega T}d\omega = \frac{T}{2\pi}\int_{-\pi/T}^{\pi/T} X^*(\omega)e^{jn\omega T}d\omega \tag{6}$$

where the auxiliary function, $X^*(\omega)$, is introduced by

$$X^*(\omega) = \frac{1}{T} \sum_{-\infty}^{\infty} X(\omega + 2\pi m/T).$$

If $|X(\omega)| = 0$ for $|\omega| \geq \pi/T$, then $X^*(\omega) = \frac{1}{T} X(\omega)$. Otherwise the values of X at $\omega + 2\pi m/T$, $m \neq 0$ also contribute to $X^*(\omega)$. (This is known as aliasing.)

Aside from changes of variables (6) corresponds to (2). Then to (1) corresponds the expression

$$X^*(\omega) = \sum_{-\infty}^{\infty} x_n e^{-j\omega nT}. \tag{7}$$

This is the discrete version of the Fourier transform (DFT). Note that it is the z-transform of $\{x_n\}$ evaluated at $z = e^{j\omega T}$. The z-transforms are discussed in the next appendix.

Finite Data

Suppose that we know $x(\cdot)$ only over $[0, NT]$ for some NT, and construct $x_{NT}(\cdot)$ to be the periodic function with period NT by repeatedly copying $x(\cdot)$ over $(-\infty, \infty)$. Thus, the Fourier series expansion

$$x_{NT}(t) = \sum_{-\infty}^{\infty} c_k(NT) e^{j2\pi kt/NT}$$

exists where

$$c_k(NT) = \frac{1}{NT} \int_0^{NT} x(t) e^{-j2\pi kt/NT} dt. \tag{8}$$

Define $\hat{X}(2\pi k/NT)$ by $NTc_k(NT)$. The original time function is constructed as in (5'):

$$x_{NT}(t) = \frac{1}{NT} \int_{-\infty}^{\infty} \hat{X}(2\pi k/NT) e^{j2\pi kt/NT}. \tag{9}$$

To (3) corresponds the Fourier transform with a finite time interval:

(10) $$\hat{X}(\omega) = \int_0^{NT} x(t)e^{-j\omega t}dt.$$

Since $x(t) = x_{NT}(t)$ for $t \leq NT$, substitute (9) into (10) and evaluate the integral

$$\hat{X}(\omega) = \int_0^{NT} \left| \frac{1}{NT} \sum_{-\infty}^{\infty} \hat{X}(2\pi k/NT)e^{j2\pi kt/NT} \right| e^{-j\omega t}dt$$

$$= \sum_{-\infty}^{\infty} \hat{X}(2\pi k/NT) \frac{1}{NT} \int_0^{NT} e^{-j(\omega-2\pi k/NT)t}dt$$

$$= \sum_{-\infty}^{\infty} \hat{X}(2\pi k/NT)e^{-j\pi NT/2} \frac{\sin \frac{\omega NT}{2}}{(\omega NT/2-\pi h)},$$

where the change of $\int$ with Σ is assumed to be legitumate.

With a sampled data (7) holds. With a finite data set (7) is replaced by

(11) $$\hat{X}^*(\omega) = \sum_0^{N-1} x_n e^{-j\omega nT}$$

where $x_n = x(nT)$.

Define X(k) by

(12) $$X(k) = \hat{X}^*(2\pi k/NT)$$

$$= \sum_0^{N-1} x_n e^{-j2\pi nk/N}.$$

This is DFT with a finite data.

To recover x_n from X(k) sequence, consider $\frac{1}{N} \sum_0^{N-1} X(k)e^{j2\pi nk/N}$. When (12) is substituted into X(k), we obtain

$$\frac{1}{N} \sum_{k=0}^{N-1} \sum_{m=0}^{N-1} x_m e^{-j2\pi mk/N} e^{j2\pi nk/N}$$

$$= \sum_{m=0}^{N-1} x_m \frac{1}{N} \sum_{k=0}^{N-1} e^{j2\pi k(n-m)/N}.$$

Using the identity

$$\frac{1}{N}\sum_{k=0}^{N-1} e^{j2\pi mk/N} = \begin{cases} 1, & m = 0 \\ 0, & m \neq 0, \end{cases}$$

the above reduces to x_n, i.e., we have established

$$x_n = \frac{1}{N}\sum_{0}^{N-1} X(k)e^{j2\pi nk/N}.$$

Now suppose $\{x_n\}$ is a mean-zero weakly stationary stochastic sequence, the variance of (11) defines

$$S(\omega) = \text{lin}\ \frac{1}{N} E|\hat{X}^*(\omega)|^2$$

$$= \sum_{-\infty}^{\infty} R_k e^{-j\omega kT}$$

as the (power) spectrum of the time series $\{x_n\}$, where $R_k = E(x_{n+k}x'_n)$ is the covariance.

A.5 The z-transform

The z-transform of a sequence $\{x_n\}$ is formally defined by

$$X(z) = \sum_{-\infty}^{\infty} x_n z^{-n}. \tag{1}$$

The one-sided z-transform is defined by $X(z) = \Sigma_{n=0}^{\infty} x_n z^{-n}$, implicitly assuming that we are interested only in that part of the sequence $\{x_n\}$ beyond the initial time n = 0. A way to recover x_k from X(z) is to calculate

$$x_k = \frac{1}{2\pi j} \oint_{|z|=1} X(z)^{k-1} dz \tag{2}$$

where the integral is carried out around the unit disc, $|z| = 1$.

A typical one-sided z-transform arises in characterizing dynamic (impulse) responses of linear systems. Although dynamic systems are described or characterized in many ways, one common way is to give a dynamic system's impulse response functions or sequence i.e., dynamic multiplier sequences, because then the (zero-state) response to any other input (exogenous) sequences is describable by

$$y(z) = H(z)U(z)$$

where

$$H(z) = \Sigma_0^{\infty} h_i z^{-i}$$

is the z-transform of the impulse responses.

Dynamic systems with rational transfer functions are stable if their poles are located in $|z| < 1$. An example of two-sided z-transforms is the covariance generating function of a weakly stationary stochastic time series. We return to these topics later.

This definition shows that the operation of forming z-transforms is linear. The z-transform of a sequence, $\{ax_n + by_n\}$, made up of a sum of scalar multiples of two other sequences $\{x_n\}$ and $\{y_n\}$ equals aX(z) + bY(z), where X(z) and Y(z) are the z-transforms of these two sequences, respectively.

Z-transforms can be discussed on at least two levels. On one level, the variable z^{-1} merely serves as a place marker in a representation of a sequence. For example, z^{-7} is associated with x_7, and serves to single out x_7 from X(z). This is certainly convenient when as infinite sum are formally put in a closed form as when $1 + z^{-1} + z^{-2} + \dots$ is represented by $1/(1 - z)$. The role of z or z^{-1} as place markers is evident in the definitions of generating functions in statistics, probability and other disciplines. On this level, we do not worry about the convergence of the formal series associated with the z-transforms. Infinite sequences are merely conveniently and compactly represented as formal power series. For example, this view is useful in relating two series that are defined by convolution:

$$c_i = \Sigma_j\, a_{i-j} b_j$$

because the z-transfrom of $\{c_j\}$, which equals A(z)B(z) where A(z) and B(z) are the z-transforms of $\{a_i\}$ and $\{b_j\}$ respectively, can be used to recover c_j.

Equation (1) shows that $z^{-1}X(z)$ corresponds to a sequence $\{y_n\}$ where $y_n = x_{n-1}$ because of the relation

$$z^{-1}X(z)z^{n-1} = X(z)z^{n-2}$$

in the integrand of (2). In other words, the multiplication by z^{-1} is a backward shift operation $z^{-1}x_n = x_{n-1}$. The lag operation L in econometrics is the same as multiplication by z^{-1}. The same holds for one-sided z-transforms.

The z-transforms of the sequence $\{h_n\}$, where $h_n = Lx_n = x_{n-1}$, then is constructed by

$$H(z) = \sum_{n=0}^{\infty} h_n z^{-n} = \sum_{n=0}^{\infty} x_{n-1} z^{-n} = x_{-1} + z^{-1} \sum_{m=0}^{\infty} x_m z^{-m} = x_{-1} + z^{-1}X(z).$$

If x_{-1} is zero, then $H(z) = z^{-1}X(z)$. Let $f_n = x_{n+1}$. Then $F(z) = \Sigma_0^\infty f_n z^{-n}$ $= \Sigma_{n=0}^\infty x_{n+1} z^{-n} = z \Sigma_{n=0}^\infty x_{n+1} z^{-(n+1)} = zX(z) - zx$. These two cases show that multiplication by z corresponds to forward shift and multiplication by z^{-1} means backward shift in the time domain.

Using one-sided z-transforms we can solve difference equations by converting them into algebraic ones just as Laplace transforms allow us to solve differential equations algebraically.

Example Given $y(z) = \Sigma_0^\infty y_n z^{-n}$, $\{y_{n+3}\}$ produces $z^3y(z) - z^3y_0 - z^2y_1 - z^3y_0$ as its z-transform.

Example The zero initial condition solution of $y_{k+n} + a_{n-1}y_{k+n-1} + \dots + a_0y_k = y_k$ has its z-transform $y(z) = G(z)/[z^n + a_n + z^{n-1} + \dots + a_0]$, where G(z) and y(k) are the (one-sided) z-transforms of $\{g_k\}$ and $\{y_k\}$ respectively.

Example Let $x_{t+1} = Ax_t + bu_t$

$$y_t = cz_t + du_t.$$

Then

$$zX(z) = AX(z) + bU(z) + zx_0$$

$$y(z) = cX(z) + dU(z).$$

Hence

$$y(z) = \{c(zI - A)^{-1}b + d\}U(z) + zc(zI - A)^{-1}x_0$$

where the first part of the z-transform of the zero-state response, and the second is that of zero-input response.

On the second and more sophisticated level, z-transforms are treated as defining analytic functions in some region of the complex plane. In some cases, the formal power series of z do converge in some region of the complex plane thus defining analytic functions. If the infinite series converge for $z = e^{j\omega T}$ with some T, then the z-transform evaluated at this z value, $X(e^{j\omega T})$, is the Fourier transform of a sampled sequence of a continuous function of time with sampling interval T.

By identifying z with $e^{j\omega T}$, we recognize (2) as the formula for the coefficient of the Fourier series expansion

$$x_k = \frac{T}{2\pi}\int_{-\pi/T}^{\pi/T} X(e^{j\omega T})e^{j\omega kT}d\omega.$$

The z-transforms are thus related to the Fourier transforms when some specific values are assigned to z. To see this we proceed as follows.

Suppose that x(t) has the Fourier transform $\tilde{X}(\omega)$ where

$$\tilde{X}(\omega) = \int_{-\infty}^{\infty} x(t)e^{-j\omega t}dt.$$

Its inverse transform is

$$x(t) = \frac{1}{2\pi}\int_{-\infty}^{\infty} \tilde{X}(\omega)e^{j\omega T}d\omega.$$

So the value of $x(\cdot)$ sampled periodically with a time interval T is given by

$$x(nT) = \frac{1}{2\pi}\int_{-\infty}^{\infty} \tilde{X}(\omega)e^{jn\omega T}d\omega.$$

Dividing the interval of integration into segments of length T each, let us rewrite the above as

$$x(nT) = \frac{T}{2\pi}\Sigma_{-\infty}^{\infty}\frac{1}{T}\int_{(2m-1)\pi/T}^{(2m+1)\pi/T} \tilde{X}(\omega)e^{j\omega nT}d\omega.$$

If we change the variable of integration from ω to $\omega' = \omega - 2\pi m/T$, we can rewrite the above as

$$x(nT) = \frac{T}{2\pi}\sum_{-\infty}^{\infty}\frac{1}{T}\int_{-\pi/T}^{\pi/T} \tilde{X}(\omega' + 2\pi m/T)e^{j\omega' nT}d\omega',$$

where we use $e^{j2\pi mn} = 1$.

Suppose the expression

$$X^*(\omega) = \frac{1}{T}\sum_{-\infty}^{\infty} X(\omega + 2\pi m/T)$$

is well defined.

Then the value of $x(\cdot)$ at $t = nT$ can be related to $X^*(\omega)$ by

(3) $$x(nT) = \frac{T}{2\pi} \int_{-\pi/T}^{\pi/T} X^*(\omega) e^{j\omega nT} d\omega.$$

The function $X^*(\omega)$ is periodic with period $2\pi/T$ because

$$X^*(\omega + 2\pi/T) = \frac{1}{T} \Sigma_{-\infty}^{\infty} \tilde{X}(\omega + 2\pi(m + 1)/T) = X^*(\omega).$$

A periodic function of t with period Ω can be represented by a Fourier series

$$x(t) = \Sigma_{-\infty}^{\infty} c_n e^{-j2\pi nt/\Omega}$$

where integration of both sides from $-\Omega/2$ to $\Omega/2$ after multiplying both sides by $e^{j2\pi mt/\Omega}$ yields

$$\int_{-\Omega/2}^{\Omega/2} x(t) e^{j2\pi mt/\Omega} dt = \Sigma_{-\infty}^{\infty} c_n \int_{-\Omega/2}^{\Omega/2} e^{j2T(m - n)t} dt = c_m \Omega$$

where we use

$$\frac{1}{\Omega} \int_{-\Omega/2}^{\Omega/2} e^{j2\pi nt/\Omega} dt = \begin{cases} 1, & n = 0 \\ 0, & n \neq 0 \end{cases}.$$

Thus

$$c_m = \frac{1}{\Omega} \int_{-\Omega/2}^{\Omega/2} x(t) e^{j2\pi mt/\Omega} dt.$$

Now fomrally represent $X^*(\omega)$ by a Fourier series. Since $X^*(\omega)$ is periodic with period $2\pi/T$, its Fourier coefficient is

(4) $$\frac{T}{2\pi} \int_{-\pi/T}^{\pi/T} X^*(\omega) e^{j2\pi mT/2\pi} d\omega = \frac{T}{2\pi} \int_{-\pi/2}^{\pi/2} X^*(\omega) e^{jm\omega T} d\omega = x(mT).$$

Now compare this with (1) to see that $x(nt)$ is the n-th Fourier coefficient of $X^*(\omega)$, i.e.,

$$X^*(\omega) = \Sigma_{-\infty}^{\infty} x(nt) e^{-jn\omega T}$$

is the Fourier series of $X^*(\omega)$.

Define a function $X(z)$ by

$$X(z) = \Sigma_{-\infty}^{\infty} x(nt) z^{-n}.$$

We recognize $X(z)$ as the z-transform of $x(\cdot)$.

Then

$$X^*(\omega) = X(e^{j\omega T}).$$

To recover x(nt) form x(z), set z to $e^{j\omega T}$ and note that $d\omega = \frac{1}{jT} dz$ to rewrite (2) as

$$x(nT) = \frac{1}{2\pi j} \int_{|z|=1} X(z) z^{n-1} dz.$$

This is the inverse z-transform given as (2).

Fourier series are also defined for functions defined on the closed interval $[-\pi, \pi]$. The Fourier coefficients are defined by

$$c_n = \frac{1}{2\pi} \int_{-\pi}^{\pi} f(x) e^{-jnx} dx, \quad n = 0, \pm 1, \ldots,$$

if the integral exists. We follow Hoffman (1962) to characterize a class of analytic functions: Let H^2 denote the class of analytic functions f in $|z| \leq 1$ for which the functions $f_r(\theta) = f(re^{i\theta})$ is bounded in L^2-norm as $r \to 1$, i.e., $\|f_r\|_2 = (\int_{-\pi}^{\pi} |f(re^{i\theta})|^2 d\theta)^{1/2}$ remains bounded as $r \to 1$.

The space H^2 then is identified with a closed subspace L^2 of the circle:

$$H^2 = \{f \in L^2 : \int_{-\pi}^{\pi} f(\theta) e^{in\theta} d\theta = 0, \ n = 1, 2, \ldots\},$$ in other words, the element $f \in H^2$ has a one-sided z-transform $f(z) = \Sigma_0^{\infty} a_n z^{-n}$, where $z = e^{-j\theta}$, because the Fourier coefficients vanish for negative integers. The shift operator T can be defined on H_2 = a space of square summable sequences of complex numbers by

$$T(a_0, a_1, a_2, \ldots) = (0, a_0, a_1, \ldots)$$

where

$$\Sigma |a_1|^2 < \infty$$

or on H^2 by

$$(Tf)(\theta) = e^{i\theta} f(\theta) \quad (= z^{-1} f(z)).$$

A.6 Some Useful Relations for Quadratic Forms

Here we collect some useful formulas involving quadratic forms. Most of results are found in Bellman [1960] but are collected here for easy reference.

1. Completion of square can be used to find the minimum and the minimizing expression in

$$\operatorname*{Min}_x (a + 2b'x + x'Qx) = a - r'Qr$$

where $Q' = Q > 0$, and $x = -Q^{-1}r$.

This follows by writing

$$a + 2b'x + x'Qx = a + (x + Q^{-1}r)'Q(x + Q^{-1}r) - r'Qr.$$

2. The extrema of $x'Qx$ subject to $x'x = 1$ is related by

$$\lambda_{min}(Q) \leq x'Qx/x'x \leq \lambda_{max}(Q),$$

or Min $x'x$ subject to $x'Qx = 1$ may analogously be phrased.

3. The minimum of $x'Qx$ subject to a linear constraint $Ax = z$ where A is an $m \times n$ matrix of rank n is achieved by $x = Q^{-1}A'(AQ^{-1}A')^{-1}z$.

4. The matrix solution of the linear differential equation

$$\dot{X} = AX + XB, \ X(0)$$

is given by $X(t) = e^{At}Ce^{Bt}$.

5. If $X = -\int_0^\infty e^{At}Ce^{Bt}dt$ exists for all C, then it is a unique solution of $AX + XB = C$. To see this, consider $\dot{Z} = AZ + ZB$, $Z(0) = C$. Assuming that $Z(t) \to 0$, as $t \to \infty$, integrate the differential equation to see that

$$-C = A(\int_0^\infty Z(s)ds) + (\int_0^\infty Z(s)ds)B.$$

6. Using the kronecher delta notation, the matrix equation $AX + XB = C$ is converted into $(A \otimes I + I \otimes B')\text{vec } X = \text{vec } C$ where $A \otimes B = (a_{ij}B)$. The matrix $A \otimes B$ has eigenvalues $\lambda_i\mu_j$ where λ_i is the i-th eigenvalue of A can μ_j is the

j-th eigenvalue of B.

The matrix equation $AX + XB = C$ has a solution for all C if and only if the eigenvalues of A do not cancel each other out, i.e., $\lambda_i + \mu_j \neq 0$ for all i and j. The uniqueness follows from linearity.

7. The algebraic equation of the form $X - AXA' = Q$ where $Q' = Q > 0$ arises in several context. It is equivalent to $(I - A \times A)\operatorname{vec} X = \operatorname{vec} Q$. The matrix $I - A \times A$ has eigenvalues $1 - \lambda_i\lambda_j$ hence $I - A \times A$ is nonsingular if and only if $\lambda_i\lambda_j \neq 1$ for all i, j. The condition $|\lambda(A)| < 1$ is sufficient.

8. Lyapunov theorem

An algebraic equation for a symmetric matrix

(1) $A'XA - X = -R$

where $R' = R > 0$ arises in may context. We call the matrix A stable if all its eigenvalues have modulus less than one. The matrix X is clearly symmetric. First, the solution matrix X is unique if A is stable. Suppose there are two solutions. The difference $X_1 - X_2$ obeys $A'(X_1 - X_2)A = (X_1 - X_2)$. To consider a simple case, suppose A has distinct eigenvalue, $Av = \lambda v$. Multiply the above by v from right, we note that

$\lambda A'(X_1 - X_2)v = (X_1 - X_2)v$, or if $X_1 - X_2 \neq 0$, then $(X_1 - X_2)v$ is an eigenvector of A' with eigenvalue $1/\lambda$. However, $|1/\lambda| > 1$ if $|\lambda| < 1$ contradicting the assumed stability of A, hence $X_1 = X_2$. The solution is unique.

By iterating (1), the solution X may be written as the sum of an infinite series

(2) $$X = R + A'RA + (A')^2RA^2 + \ldots.$$

This is well defined because A is stable. To see this we need only to note

$$\bar{v}'\{\sum_{h=0}^{\infty} (A')^h R A^h\} v = (\bar{v}'Rv) \sum_{0}^{\infty} |\lambda|^{2h} = (\bar{v}'Rv)/(1-|\lambda|^2) < \infty$$

for any eigenvector v of A. This argument establishes finiteness of (2) when A has distinct eigenvalue and that X is positive definite. Even with some repeated eigenvalues, the infinite sum can be shown to be bounded and X be positive definite.

The converse also true because $\bar{v}'(A'XA - X)v = (|\lambda|^2 - 1)\bar{v}Xv = -\bar{v}'Rv < 0$ implies that $|\lambda| < 1$, i.e., A is stable.

These are summarized as <u>Lyapunov Theorem.</u> The matrix A is stable it and only if (1) has a unique symetric positive definite solution.

9. The integral $J = \int_0^\infty (X'BX)dt$ when evaluated along a solution of $x' = Ax$ equals $-x(0)'Yx(0)$, where $A'Y + YA = B$ or $Y = -\int_0^\infty e^{A't}Be^{At}dt$. This can be seen by integrating $\frac{d}{dt}(x'Yx) = x'Bx$.

A.7 Calculation of the Inverse, $(zI - A)^{-1}$

A recursive procedure for calculating $(zI - A)^{-1}$ is available: See Aoki [1976; p.45].

$$(zI - A)^{-1} = \frac{1}{d(z)} [z^{n-1} + B_1 z^{n-2} + \ldots + B_{n-1}]$$

where

$$d(z) = |zI - A| = z^n + a_{n-1} z^{n-1} + \ldots + a_0$$

and

$$B_1 = A + a_{n-1} I$$

$$\cdot \; \cdot \; \cdot \; \cdot$$

$$B_\ell = AB_{\ell-1} + a_{n-\ell} I \qquad \ell = 2 \ldots n-1$$

$$0 = AB_{n-1} + a_0 I.$$

When this algorithm is applied to a single-input-single-output system (A, b, c) in the phase canonical form we can readily establish that $B_i b = e_{n-i}$, $i = 1 \ldots n-1$. Hence we can write the transfer function as

$$c(zI - A)^{-1} b = (c_{n-\ell} z^{n-\ell} + \ldots + c_0)/d(z).$$

A.8 Sensitivity Analysis of Optimal Solutions: Scalar-Valued Case

Asymptotic behavior of optimally controlled system state vector is the same, i.e., the state vector approaches zero for any choice of weighting matrices Q and R provided that the dynamics are controllable and (A, H) is detectable, where $Q = H'H$. The latter condition depends on which components of the state vector are actually included in the performance indices.

Dynamic behavior i.e., the manner of approaching zero, however, are influenced by our choice of Q and R. The transient response of the dynamics with the optimal control is determined by the eigenvalues of the matrix $(A - BK^*)$ where $K^* = R^{-1}B'p^*$ and P^* is the solution of the matrix Riccati equation.

We now conduct a kind of root-locus analysis for a special class of dynamics. See Aoki [1976, 1981] for some description on the root-locus method. We limit ourselves to problems with scalar-valued decision variables and scalar-valued data and correspondingly specialize R to a scalar r and B to a vector b, and $Q = hh'$ where h is a column vector. We follow Kailath [1980] in our development. The transfer function between u and $y = h'x$ is then

(1) $$h'(sI - A)^{-1}b = n(s)/d(s).$$

Here $d(s) = |sI - A|$ is the characteristic polynomial of A and where we assume that d and n have no common factors. (This follows if the dynamics are controllable and observable as we have assumed.)

The Riccati equation becomes $A'P + PA - Pbb'P/r + hh' = 0$, and the optimal feedback gain $-b'p/r$. First, rearrange the Riccati equation by adding and subtracting sP, where s is the Laplace transform variable, to read

$$P(sI - A) + (-sI - A')P + Pr^{-1}b'P = hh'.$$

Multiply the above from left by $b'(-sI - A')^{-1}$ and from the right by $(sI - A)^{-1}b$ to rewrite the above as

$$(*) \qquad b'(-sI - A')^{-1}k + k'(sI - A)^{-1}b + b'(-sI - A')^{-1}kk'(sI - A)b = r^{-1}b'(-sI - A')^{-1}hh'(sI - A)^{-1}b.$$

Let the characteristic polynomial of the closed loop system $|sI - A + bk'|$ be written as

$$d_k(s) = d(s)\{1 + k'(sI - A)^{-1}b\}$$

where we use the identity $|1 + cd'| = 1 + d'c$. See Aoki [1976; Appendix B]. The zeros of this polynomial are the eigenvalues of the dynamics with feedback gain vector k.

Then (*) is used to simplify the expression

$$d_k(-s)d_k(s) = d(-s)d(s)\{1 + r^{-1}b'(-sI - A')^{-1}hh'(sI - A)^{-1}b\} = d(-s)d(s) + r^{-1}n(s)n(-s)$$

where we use $b'(-sI - A')^{-1}h = h'(-sI - A)^{-1}b = n(-s)$.

Next we claim that the eigenvalues of $(A - br^{-1}b'P)$ which determine the transient response of the optimally controlled dynamics are the n stable roots of the polynomial $d_k(-s)d_k(s)$, where n is the degree of the polynomial d(s), i.e., the parametric dependence of eigenvalues or r is exhibited by*

$$(2) \qquad d(-s)d(s) + r^{-1}n(-s)n(s) = 0.$$

This gives a generalized sort of the root-locus plot for eigenvalues. For

---

* From the Hamiltonian formulation, the state vector z augumented by the adjoint vector λ is governed by

$$\frac{d}{dt}\binom{z}{\lambda} = M\binom{z}{\lambda} \qquad \text{where} \qquad M = \begin{pmatrix} A & -br^{-1}b' \\ -hh' & -A \end{pmatrix}.$$

Note that $|\lambda I - M| = d_k(s)d_k(-s)$.

large r, i.e, with heavy cost of control, the transient responses are governed by eigenvalues which are approximately equal to the roots of d(s) = 0, i.e., the controlled system eigenvalues are near those of the uncotrolled system and the system behaves almost as the uncontrolled dynamics itself. With r approaching zero, q of the eigenvalues are given by the roots of

$$n(s) = 0$$

where q = deg n. These are the zeros of the uncontrolled system transfer function (1). The coefficients of the polynomial n(•) can be recursively determined by Leverrier's method (Aoki [1976; p.45]), for example. The remaining (n - m) eigenvalues approaches ∞ along some asymptotes. These can be determined by retaining the largest term in $|s|$ in (2):

$$(-1)^n s^{2n} + (-1)^m r^{-1} n_0^2 s^{2m} = 0$$

where n_0 is the coefficient of s^m in n(s).
The (n - m) asymptotes that lie in the left half s-place are thus (n - m) stable branches of

$$s^{2(n-m)} = (-1)^{n-m-1}(n_0^2/r) \qquad \text{and} \qquad r \to \infty.$$

A.9 Common Factor in ARMA Model and Controllability

This appendix establishes a connection between controllability and presence of common factors in ARMA models. To this end, let ε_t be a scalar, and b, c, and d be vectors.

A dynamic system represented as

$$(S)\begin{cases} x_{t+1} = Ax_t + b\varepsilon_t \\ y_r = cx_t + d\varepsilon_t \end{cases}$$

has the transfer function

$$H(z) = d + c(zI - A)^{-1}b = \psi(z)/\phi(z)$$

where

$$\phi(z) = |zI - A|,$$

and

$$\psi(z) = \begin{vmatrix} ZI - A & b \\ -c & d \end{vmatrix},$$

hence can be put into an ARMA model form

$$\phi(L)y_t = \psi(L)\varepsilon_t.$$

In the above, a matrix identity*

$$\begin{pmatrix} I & 0 \\ c(zI - A)^{-1} & 1 \end{pmatrix} \begin{pmatrix} zI - A & b \\ -c & d \end{pmatrix} = \begin{pmatrix} zI - A & b \\ 0 & H(z) \end{pmatrix}$$

and the corresponding determinantial equality are used to show that $\phi(z) \cdot H(z)$ equals $\psi(z)$.

We now show that if $\phi(\cdot)$ and $\psi(\cdot)$ has a common factor then (S) is either not controllable or observable. Suppose $\phi(z_1) = 0$ and $\psi(z_1) = 0$ where z_1 is one of the eigenvalues of A. (From $\phi(z) = |zI - A|$, the roots of $\phi(\cdot)$ are all eigenvalues of A.) Vanishing $\psi(z_1)$ implies that there exists a vector (ξ, η) not identically zero such that

---

* Or more directly, use the matrix identity

$$\begin{pmatrix} zI - A & b \\ -c & d \end{pmatrix} = |zI - A| \cdot (d + c(zI - A)^{-1}b).$$

$$\begin{pmatrix} zI - A & b \\ -c & d \end{pmatrix}\begin{pmatrix} \xi \\ \eta \end{pmatrix} = 0.$$

If η is zero, then $\xi \neq 0$ and $(z_1I - A)\xi = 0$ and $c\xi = 0$ or $\xi'[c', A'c', (A')^{n-1}c) = 0$, with $\xi \neq 0$. This means that the system (S) is not observable. If $\eta \neq 0$, then $(z_1I - A)\xi + b\eta = 0$. Let $\phi(z) = (z - z_1)\tilde{\phi}(z)$. On multiplying $\tilde{\phi}(z)$ from the left, this equation becomes $\tilde{\phi}(z)b\eta = 0$ or because $\phi(A)$ vanishes by the Cayley-Hamilton theorem, we have $\tilde{\phi}(A)b = 0$, i.e., (A, b) is not a controllable pair.

The converse is also straigtforward to show. (See Kailath [1980] or Chen [1970] for example.)

A.10 Non-Controllability and Singular Probability Distribution

If a Markovian model generating a time series $\{y_t\}$ is not controllable, then the probability distribution of y becomes singular. To see this simply, let a p-dimensional y_t be generated by a stable dynamics

$$y_{t+1} = Ay_t + b\varepsilon_{t+1}, \qquad y_0 = 0$$

where $\dim[b, Ab, \ldots, A^{p-1}b] < p$, and $\{\varepsilon_t\}$ is the usual mean-zero white noise sequence.

By this non-controllability assumption, there is a p-vector α such that $\alpha'[b, Ab, \ldots, A^{p-1}b] = 0$. Becuase $y_t = \Sigma_{s=0}^{t} A^s b\varepsilon_{t-s}$ by the dynamics, we note that $E\{\alpha' y_t \varepsilon_{t-s}\} = 0$ for all $s \geq 0$. By the Cayley-Hamilton theorem $\alpha'\mathfrak{C} = 0$ for $\mathfrak{C} = [b, Ab, \ldots,]$.

Thus, $\alpha' y_t$ equals zero in the mean square sense. The distribution of y's is thus confined to some subspace in the space of all mean-zero, finite variance random variables. This shows up as the rank of the (sample) covariance matrix of $\{y_t\}$ being less than the dimension of the vector y.

A non-controllable dynamics contain a controllable subsystem by an appropriate partition of the vector y. This subvector may also be identified by a suitable partition of the covariance matrix of y. Let $\Sigma = \text{cov}(y)$ and denote by Γ and Λ the matrices of eigenvectors and eigen values respectively; $\Sigma\Gamma = \Gamma\Lambda$ where $\Lambda = \text{diag}(\lambda_1, \ldots, \lambda_q, 0\text{--}0)$, r with $q > p$.

The characteristic function for y is $E(e^{it'y}) = \exp -\frac{1}{2} t'\Sigma t$. Let $t = \Gamma\theta$ and denote $\Gamma' y$ by ν. Then $E(e^{it'y}) = E(e^{i\theta'\Gamma' y}) = E(e^{i\theta'\nu}) =$ $\exp -\frac{1}{2}\theta'\Gamma'\Sigma\Gamma\theta = e^{-\frac{1}{2}\theta'\Lambda\theta} = \exp -\frac{1}{2}\Sigma_{i=1}^{q}\lambda_i\theta_i^2$, showing that $\nu_1, \ldots, \nu_q$ are independently distributed and $\nu_j = 0$, $j = q+1, \ldots, p$ with probability 1 because their characteristic function is 1. (See Aoki: [1967; Appendix III].)

A.11 Spectral Decomposition Representation

Suppose an $n \times n$ matrix A has n linearly independent vectors $\{u_i\}$ and n (not necessarily distinct) eigenvalues counting multiplicities. For example $A = \begin{pmatrix} 2 & 0 \\ 0 & 2 \end{pmatrix}$ has $\begin{pmatrix} 1 \\ 0 \end{pmatrix}$ and $\begin{pmatrix} 0 \\ 1 \end{pmatrix}$ as two linearly independent eigenvectors with eigenvalues $\lambda_1 = \lambda_2 = 2$.

Let $U = [u_1', \ldots, u_n]$ and $V = U^{-1}$ and write the row vectors of V as v_i', $i = 1, \ldots, n$. The vector u_i is the right eigenvector while v_i' is called the left eigenvector of λ_i because $v_i'A = \lambda_i v_i'$.

By definition

$$AU = U\Lambda$$

where $$\Lambda = \mathrm{diag}(\lambda_1, \ldots, \lambda_n).$$

Thus we obtain the sepctral decomposition of the matrix A

$$A = U\Lambda V = \Sigma_{i=1}^{n} \lambda_i u_i v_i'.$$

This representation is useful in evaluating dynamic effects because it effectively represents dynamics as a parallel array of n scalar dynamics. To illustrate, note that

$$e^{At} = \Sigma_{i=1}^{n} e^{\lambda_i t} u_i v_i'.$$

Therefore the effect of a scalar exogenous variable u on y where

$$y = c'x$$

$$\dot{x} = Ax + bn, \quad x(0) = 0$$

can be decomposed into n components

$$y(t) = \Sigma_{i=1}^{n} e^{\lambda_i t} (c'u_i)(v_i'b) \int_0^t e^{-\lambda_i \tau} u(\tau) d\tau,$$

showing that if c is orthogonal to u_i, then y(t) does not contain a component proportional to $e^{\lambda_i t}$ (i.e., the i-th mode is not observed by y) and that if $v_i'b$ is zero u does not influence the i-th mode, i.e., the i-th mode is not controllable.

For other illustration of such model decompostion see Aoki [1964 & 1968].

A.12 Singular Value Decomposition Theorem

Effects of small perturbations are essential parts of many analysis related to stability, optimality and parameter sensitivity. Dynamic systems are assessed for their structural stability and parameter sensitivity as well as variational dynamics are used to study "neighboring" time paths of solutions. Solutions of algebraic equations (overdetermined or otherwise) are incomplete unless some "condition numbers" are calculated to indicate degree of robustness of solutions or ill-poseduen of problem formulation. Similarly, computational error analysis is a must in evaluating algorithms in numerical analysis. In statistics, principal component analysis, canonical correlation analysis and the like exist to perform similar functions.

Here we examine singular value decomposition as a tool for unifying sensitivity analysis in some time series analysis, as well as a practical way for determining ranks of numerically determined matrices. The fact that any $(m \times n)$ matrix A is expressible as $A = U\Sigma V^*$ where $U^*U = I_m$, $V^*V = I_n$ and $\Sigma = \text{diag}(\Sigma_\nu, 0)$, $\Sigma_\nu = \text{diag}(\sigma_1, \ldots, \sigma_r)$ where $r = \text{rank } A$, is known as the singular value decomposition theorem. (See Strang [1973] or Golub and Reinsch [1970] for example. The proof is summarized in Appendix. Using this decomposition we can easily establish some properties of rectangular matrices (see Appendix for proof)

(i) $$A^* = V\Sigma' U^*$$

(ii) $$A^*A = V\Sigma'\Sigma V^*, \quad AA^* = U\Sigma\Sigma' U^*$$

(iii) Let A, B be $(m \times n)$ matrices. Then

$$\|A - B\| = \max_{x \neq 0} \|(A - B)x\| / \|x\|$$

$$= \text{The largest singular value of } (A - B).$$

(iv) $$A^+ = V\Sigma^+U^* \text{ where } \Sigma^+ = \text{diag}(\Sigma_r^{-1}, 0)$$

is the Moore-Penrose pseudo-inverse. (See Aoki [1967], Appendix II also.)

(v) The condition number of A is σ_1/σ_n.

(vi) Let A be $n \times n$ with eigenvalues $\lambda_1, \ldots, \lambda_n$ arranged in the order of decreasing magnitude. Then $\sigma_1 \geq |\lambda_i| \geq \sigma_n$ and $\text{cond}(A) \geq |\lambda_1/\lambda_n|$.

An easy application is to the sensitivity analysis of solutions of algebraic equations. Let $Ax = b$ where A is $n \times n$. A slight error in specifying b produces an error in $x : A(x + \Delta x) = b + \Delta b$ or $A\Delta x = \Delta b$. Using the singular value decomposition $A = U\Sigma V^*$, define $\Delta b = U\Delta\tilde{b}$ and $\Delta x = V\Delta\tilde{x}$. Then $\sigma_i\tilde{x}_i = \tilde{b}_i$ and $\sigma_i\Delta\tilde{x}_i = \Delta\tilde{b}$. From $\sigma_n \leq \|b\|/\|x\| \leq \sigma_1$ and $\sigma_n \leq \|\Delta b\|/\|\Delta x\| \leq \sigma_1$, we can bound

$$(\text{cond}(A))^{-1} \leq \frac{\|\Delta x\|/\|x\|}{\Delta b/\|b\|} \leq \text{cond}(A).$$

Suppose now that A is $m \times n$ where $\text{rank}(A) = r$. The singular value decomposition with $b = U\tilde{b}$ and $y = V\tilde{x}$ shows that

$$\sigma_i\tilde{x}_i = \tilde{b}_i \qquad i = 1, \ldots, r$$

and

$$0 = \tilde{b}_i \qquad i = r+1, \ldots, n.$$

The solution, then, $\tilde{x}_i = \tilde{b}_i/\sigma_i$, $1 \leq 1' \leq r$ and $\tilde{x}_i$ is undetermined for 0, $i > r+1$.

From $\|b - Ax\|^2 = \Sigma_{i=1}|\tilde{b}_i - \sigma_i\tilde{x}_i|^2 + \Sigma_{r+1}^m|\tilde{h}_i|^2 \geq \Sigma_{r+1}^m|\tilde{h}_i|^2$, $\tilde{x}_i = \tilde{b}_i/\sigma_i$ is the least square solution and $\tilde{x}_i = 0$, $i \geq r+1$ produces the minimum norm solution, i.e., $x = A^+b$.

A.13 Hankel Matrices

Here, we cite several problems in which Hankel matrices appear as a part of the problem descriptions or solutions.

A deterministic counterpart of the prediction problem is to calculate future values from past input sequences i.e., assuming no more inputs $u_{t+1}, \ldots,$. We note that for the model (2) of Chapter 7

$$\begin{pmatrix} y_{N+1} \\ y_{t+2} \\ \cdot \\ \cdot \\ \cdot \\ y_{t+N} \end{pmatrix} = \begin{pmatrix} C \\ CA \\ \cdot \\ \cdot \\ \cdot \\ CA^{N-1} \end{pmatrix} \chi_{N+1}$$

where χ_{t+1} is related to the current and past inputs with zero initial conditions by

$$\chi_{N+1} = [B, AB, \ldots, A^N B] \begin{pmatrix} u_N \\ u_{N-1} \\ \cdot \\ \cdot \\ \cdot \\ u_0 \end{pmatrix}.$$

Together, future observations $y_{t+1}, \ldots, y_{t+N}$ is related to the current and past u's, $u_0, u_1 \ldots u_N$, by

$$\begin{pmatrix} y_{t+1} \\ \cdot \\ \cdot \\ \cdot \\ y_{t+N} \end{pmatrix} = H_n \begin{pmatrix} u_N \\ \cdot \\ \cdot \\ \cdot \\ u_0 \end{pmatrix}.$$

The same matrix $CA^{i-1}B$ as in (4) of Chapter 7 appear when the transfer matrix $C(zI-A)^{-1}B$ is expanded into Laurent series

$$C(zI-A)^{-1}B = CB + CAB/z + CA^2B/z^2 \ldots$$

Identifiability

Hankel matrices with auto-correlation coefficient as elements arise in some identification problems. The next ARMA model illustrates. Consider a scalar y_t and ε_t related by

$$(*) \qquad \{1+\alpha(L)\}y_t = \{1+\beta(L)\}\varepsilon_t$$

where

$$\alpha(L) = \sum_{i=1}^{p} \alpha_i L^i \text{ and } \beta(L) = \sum_{i=1}^{q} \beta_i L^i, \; q \leq p.$$

The unknown parameters are collected into a vector θ, $\theta = (\alpha_1, \ldots, \alpha_p, \beta_1, \ldots, \beta_q)'$. The output auto-correlation is denoted by $R_i(\theta)$ where $E(y_t y_{t-h}) = R_h(\theta)$. Now calculate the covariances of (*) with y_{t-j} for $j \geq p$ to obtain

$$[R_\tau R_{\tau+1}, \ldots, R_{p+\tau}] \begin{pmatrix} \alpha_p \\ \cdot \\ \cdot \\ \cdot \\ \alpha_1 \end{pmatrix} = -R_{p+\tau}, \; \tau = 1, \ldots, p.$$

These equations can be arranged as

$$H_p(\theta) a_p(\theta) = r_p(\theta)$$

where

$$H_p(\theta) = \begin{pmatrix} R_1(\theta), & \ldots, & R_p(\theta) \\ \cdot & & \\ \cdot & & \\ \cdot & & \\ R_p(\theta), & \ldots, & R_{2p-1}(\theta) \end{pmatrix}, \; a_p(\theta) = \begin{pmatrix} \alpha_p \\ \cdot \\ \cdot \\ \cdot \\ \alpha_1 \end{pmatrix}$$

and

$$r_p(\theta) = (R_{p+1}(\theta), \ldots, R_{2p}(\theta))'.$$

The matrix $H_p(\theta)$ is a Hankel matrix. The parameters $\beta_1, \ldots, \beta_q$, $q \leq p$, are related to $a_p(\theta)$ by another matrix which is a Toeplitz matrix

$$\begin{pmatrix}\beta_q\\ \cdot\\ \cdot\\ \cdot\\ \beta_1\end{pmatrix} = \begin{pmatrix} R_0, & R_1, & \dots, & R_p\\ R_{-1}, & R_0, & \dots, & R_{p-1}\\ R_{q+1}, & \dots\dots, & & R_{p+q+1}\end{pmatrix}\begin{pmatrix}\alpha_p\\ \cdot\\ \cdot\\ \alpha_1\\ 1\end{pmatrix}.$$

We note from (*)

$$y_{t+p+j} + \alpha_1 y_{t+p+j-1} + \dots + \alpha_p y_{t+j} = \beta_0 \varepsilon_{t+p+j} + \beta_1 \varepsilon_{t+p+j-1} + \dots + \beta_p \varepsilon_{t+pj}.$$

Hence for $j \geq 0$, the right hand side is independent of ε_t, ε_{t-1}, ...,. Therefore, the conditional predictions of future observations satisfy a linear relation

$$y_{t+p+j|t-1} + \alpha_1 y_{t+p+j-1(t+1)} + \dots + \alpha_p y_{t+j(t-1)} = 0 \quad \forall j \geq 0,$$

showing that the space of predictors related by the Hankel matrix or the rows of $\mathcal{H}$ as in the first example of Hankel matrices eventually become dependent on previous rows. Actually p is the smallest integer such that $y_{t+p|t-1}$ is linearly depended on its predecessors $y_{t+i|t-1}$, $i = 0, \dots, p-1$. This observation is important because it generalizes to a vector-valued process and gives a constructive procedure for Markov models.

Hankel matrices arise in yet another way in approximating a given impulse response sequence by that of a dynamics with a rational transfer function. Although the next example is not the way lower-order models are constructed (this method suffers from numerical instability), its simplicity conveys the idea of approximation well.

Example — Padé Approximation of a Transfer Function

Suppose a rational approximation b(z)/a(z) is desired to a given impulse sequence $\{h_i\}_{i=0}^n$ where $a(z) = 1 + a_1 z^{-1} +, \dots, + a_n z^{-n}$ and $b(z) = b_0 + b_1 z^{-1} +, \dots, + b_n z^{-n}$. One way is to choose a's and b's so that the first (n+1) elements of

$$\tilde{H}(z) = b(z)/a(z) = \tilde{h}_0 + \tilde{h}_1 z^{-1} +, \ldots, + \tilde{h}_n z^{-n} + \ldots,$$

exactly match the given sequence. This approximation H(z) to $\{h_i\}$ is known as Padé approximation. Equating the coefficient of $z^{-(n+1)}, z^{-(n+2)}, \ldots, z^{-2n}$ respectively, a's must satisfy

$$\begin{pmatrix} h_1, & \ldots, & h_n \\ \cdot & & \\ \cdot & & \\ \cdot & & \\ h_n, & \ldots, & h_{2n-1} \end{pmatrix} \begin{pmatrix} a_n \\ a_{n-1} \\ \cdot \\ \cdot \\ a_1 \end{pmatrix} = - \begin{pmatrix} h_{n+1} \\ \cdot \\ \cdot \\ \cdot \\ h_{2n} \end{pmatrix}.$$

where the Hankel matrix again appears and

$$\begin{pmatrix} b_n \\ b_{n-1} \\ \cdot \\ \cdot \\ \cdot \\ b_0 \end{pmatrix} = \begin{pmatrix} h_0, & h_1, & \ldots, & h_n \\ 0, & h_0, & \ldots, & h_{n-1} \\ 0, & 0, & h_0, \ldots, & h_{n-2} \\ & & & \\ 0, & \ldots, & \ldots, & h_0 \end{pmatrix} \begin{pmatrix} a_n \\ a_{n-1} \\ \cdot \\ \cdot \\ a_1 \\ 1 \end{pmatrix}.$$

Phase-Canonical Transformation

A somewhat more technical use of a Hankel matrix arises in transforming a dynamic (controllable) system into a phase canonical form.

Consider a single-input-single-output Markov model

$$\chi_{t+1} = A\chi_t + bu_t$$

$$y_t = c\chi_t$$

where

$$A = J - e_n a', \; a' = (a_0, a_1, \ldots, a_n), \; b' = e'_n = (0 \ldots 01),$$

$$c = (c_0, c_1, \ldots c_{n-\ell}, 0 \ldots 0)$$

with

$$J = \begin{pmatrix} 0 & 1 & 0 & \ldots & 0 \\ 0 & 0 & 1 & \ldots & 0 \\ 0 & \ldots & \ldots & 0 & 1 \\ 0 & \ldots & \ldots & \ldots & 0 \end{pmatrix}.$$

The matrix J is a shift matrix. For example $Je_n = e_{n-1}$, where $e_{n-1} = (0\ldots 010)$, or $Ja = (a_1, \ldots, a_{n-1}, 0)$.

A dynamic system

$$\zeta_{t+1} = F\zeta_t + gu_t$$

$$y_t = d\zeta_t$$

Can be put into the phase canonical form by a nonsingular transformation T, i.e.,

$$A = T^{-1}FT,\ b = T^{-1}g,\ c = dT$$

where

$$[g, Fg, \ldots, F^{n-1}g] = TH^{-1}$$

where

$$H^{-1} = [b, Ab, \ldots, A^{n-1}b]$$

or the required transformation is

$$T = [g, Fg, \ldots, F^{n-1}g]H.$$

This matrix H turns out to be a Hankel matrix

$$H = \begin{pmatrix} a_1 & a_2 & \cdots & a_{n-1} & 1 \\ a_2 & a_3 & & 1 & 0 \\ \cdot & \cdot & & 0 & \cdot \\ \cdot & a_{n-1} & & \cdot & \cdot \\ a_{n-1} & 1 & & \cdot & \cdot \\ 1 & 0 & \cdots & 0 & 0 \end{pmatrix}.$$

Toeplitz matrices

Hankel matrices are related to Toeplitz matrices by simple transformations. Listing the columns in reverse order of a Toeplitz matrix

$$T = \begin{pmatrix} R_0 & R_{-1} & \cdots & R_{-p+1} \\ R_1 & \ddots & & \\ \vdots & & \ddots & \\ R_{p-1} & \cdots & & R_0 \end{pmatrix}$$

produces a Hankel matrix

$$H_c = \begin{pmatrix} R_{-p+1} & \cdots & R_0 \\ \vdots & ⋰ & R_1 \\ \vdots & ⋰ & \\ R_0 & & R_{p-1} \end{pmatrix}.$$

Let J denote a p by p matrix with ones along the counter-diagonal line and zero everywhere else,

$$J = \begin{pmatrix} 0 & \cdots & 0 & 1 \\ 0 & \cdots & 1 & 0 \\ \vdots & & & \\ 1 & \cdots & 0 & 0 \end{pmatrix}.$$

This matrix is symmetric and idempotent, i.e., $J^2 = I$. Using J the above operation can be expressed as $H = TJ$.

Rearrangement of the rows of T in the reverse order also results in a different Hankel matrix. This matrix is related to T by JT:

$$H_r = \begin{pmatrix} R_{p-1} & \cdots & R_0 \\ \vdots & & \\ R_0 & \cdots & R_{-p+1} \end{pmatrix}.$$

Clearly, Toeplitz matrices can be obtained by pre- or post-multiplication of a given Hankel matrix by the matrix J as well. By calculating the singular value decomposition of T and H, we see that both matrices have the same singular values.

Elsewhere in this lecture notes of a time series, we calculate sample covariance matrices and arrange them to form a Hankel matrix

$$H = \begin{bmatrix} R_1 & R_2 & \cdots\cdots & R_p \\ R_2 & R_3 & \cdots\cdots & R_{p+1} \\ \vdots & & & \\ R_p & \cdots\cdots\cdots & & R_{2p} \end{bmatrix}.$$

Reversing the order of the columns results in the Toeplitz matrix

$$T_c = \begin{bmatrix} R_p & R_{p-1} & \cdots\cdots & R_1 \\ \vdots & & & \\ R_{2p} & \cdots\cdots\cdots & & R_p \end{bmatrix}$$

which is the south-west corner p by p submatrix of the 2p by 2p covariance matrix

$$T(2p) = \begin{bmatrix} R_0 & R_1 & \cdots\cdots & R_{-2p+1} \\ \vdots & & & \vdots \\ R_{p-1} & & & \vdots \\ R_p & & & \vdots \\ \vdots & & & \vdots \\ R_{2p-1} & & & R_0 \end{bmatrix}.$$

Reversing the rows of H yields the north-east corner of T(2p) if the time series is stationary.

Thus, a Gram-Schmidt orthogonalization works with the p by p main principal submatrix of T(2p), and the singular value decomposition of the Hankel matrix made up of covariance matrices basically works with the p by p off-diagonal submatrix.

A.14 Dual Relations*

Duality concepts or dual relations are important in static optimization such as linear, nonlinear programming and dynamic programming. See Aoki [1971; Chapt.6], Whittle [1982; Chapt.16] and numerous other books on linear and nonlinear programming.

Here we consider dual relations that arise in intertemporal optimization i.e., optimization over time. Comparison of the two sets of recursive relations that arise in optimal filtering and regulation of linear dynamic systems or the corresponding Ricatti equations reveals a remarkable resemblance. Actually, a one-to-one correspondence can be established for various terms in the recursions for these two classes of optimization problems, as we shortly demonstrate. Calculations of Kalman filter gains, and conditional variances of the estimation errors are dual in this sense to those of feedback gains and the calculations of the so-called cost-to-go in the regulator problems.

To be more precise, recall that we have established elsewhere that the minimum of a quadratic cost (so-called cost-to-go)

$$J_{t,N} = \sum_{k=t}^{N} (y_k' V y_k + d_k' T d_k)$$

subject to linear dynamics

$$\chi_{k+1} = F\chi_k + Gd_k,$$

$$y_k = H\chi_k$$

is achieved by a linear reaction rule

$$d_k = -\Gamma_k \chi_k$$

and the cost is a quadratic fundation of χ_t

---

* This section is based in part on Aoki [1967].

$$J_{t,N} = \chi_t' S_{N-t} \chi_t$$

where

$$\Gamma_{n-k} = (T + G'S_{N-k-1}G)^{-1} S_{N-k-1}F.$$

Here the Riccati equation obeys

$$S_{N-t} = F'S_{N-t-1}F + H'VH - F'S_{N-t-1}G(T + H'_{N-t-1}H)^{-1}G'S_{N-t-1}F$$

with statistics

$$E\begin{pmatrix} u_k \\ v_k \end{pmatrix} = \begin{pmatrix} 0 \\ 0 \end{pmatrix}, \quad \text{cov}\begin{pmatrix} u_k \\ v_k \end{pmatrix} = \begin{pmatrix} Q_k & 0 \\ 0 & R \end{pmatrix}.$$

The filter gain is given by

$$\Lambda_t = AP_tC'(R + CP_tC')^{-1},$$

where the relevant Riccati equation obeys

$$P_{t+1} = AP_tA' + BQB' - AP_tC'(R + CP_tC')^{-1}CP_tA',$$

where P_t is the error covariance matrix

$$P_t = \text{cov}(\chi_t - \chi_{t|t-1}).$$

Comparison of the above yields the following correspondence

Filter		Regulation
A	$\longleftrightarrow$	F'
B	$\longleftrightarrow$	H'
C	$\longleftrightarrow$	G'
R	$\longleftrightarrow$	T
Q	$\longleftrightarrow$	V
P_{t+1}	$\longleftrightarrow$	S_{N-t}
Λ_{t+1}	$\longleftrightarrow$	Γ_{N-t}

Aside from the fact that time indexing of P, S, Λ and Γ move in the opposite direction i.e., the indexing of the filtering problem goes forward in time but that of regulator is in terms of the time to the end of the planning horizon, (this is why S_{N-t} if indexed as S_t then t is time to go and $\chi_t' S_t \chi_t$ is the cost-to-go, i.e., the cost incurred in the last t periods of regulation), the first three correspondence reveals that the controllability criterion and the controllability Grammian is dual to that of the observability criterion and the observability Grammian.

A.15 Quadratic Regulation Problem: Continuous Time Systems

The question of how to contorl or regulate linear dynamic systems to minimize quadratic costs or maximize quadratic performance indices has been rather fully explored using several alternative approaches. We will follow two such strands; one based on a simple matrix relationship as noted in Bellman [1960; p.175], and the other by Dynamic Programing which is also developed by Bellman [1957].

Suppose that the integral

(1) $$J = \int_0^\infty x'(t)Cx(t)dt$$

where

$$\dot{x}(t) = Fx(t)$$

if finite where A and C are constant n × n matrices. One way to represent J is to look for a constant matrix P such that

$$\frac{d}{dt}\{x'(t)Px(t)\} = -x'(t)Cx(t)$$

because if a (unique) P is found and if $x'(t)Px(t)$ goes to zero as t approaches infinity, then integrating the above we can set

(2) $$J = x'(0)Px(0).$$

The matrix P satisfying

$$F'P + PF = -C$$

evidently meets our requirement.

Now extend the problem from evaluating a quadratic expression such as J to minimizing one:

(3) $$J = \int_0^\infty \{x'(t)Qx(t) + u'(t)Ru(t)\}dt$$

where Q and R are constant matrices $Q \geqslant 0$ and $R > 0$,

where the dynamics is now given by

(4) $$\dot{x} = Ax + Bu.$$

Suppose we restrict u to be of constant feedback type

(5) $$u(t) = -Kx(t).$$

This does not constrain our search for optimal control because the optimal control is of this type, as we later show.

With (5), (3) and (4) are transformed into

$$J = \int_0^\infty x'(t)(Q + K'RK)x(t)dt$$

where

$$\dot{x}(t) = (A - BK)x(t).$$

Our result (2) then states that

$$J = x'(0)Px(0)$$

where P is a solution of

(6) $$(A - BK)'P + P(A - BK) = -(Q + K'RK).$$

The best choice of K, then, must be the one that minimizes J with respect to K. Assuming the existence of such K and the corresponding P, denote them by K* and P* respectively. Necessarily the first order variation of P* at P* is zero in response to a small deviation of K from K*, i.e.,

$\Delta P^* = 0$ in response to a "small" $\Delta K \neq 0$.

Taking the variation of (6) yields then

$$\Delta K'(-B'P^* + RK^*) + (-B^*P^* + RK^*)'\Delta K = 0.$$

Because ΔK is arbitrary, the necessary condition for K* and P* is $RK^* = B'P^*$ or

(7) $$K^* = R^{-1}B'P^*$$

substituting (7) into (6), the matrix P* corresponding to this optimal gain is determined by an algebraic relation

(8) $$A'P^* + P^*A - P^*BR^{-1}B'P^* + Q = 0.$$

This equation is known as an algebraic Riccati equation.

Equation (7) can also be shown to be sufficient because of the quadratic nature of our criterion function.

We later show that the existence of P*, based on the assumed finiteness of the expression J, is guaranteed if (A, B) is a controllable pair. Finiteness of J implies that $x'Qx + u'Ru$ converges to zero as t goes to infinity. To ensure that x(t) itself goes to zero, we need an additional assumption that (A, H) is an observable pair whose H is a factor such that $Q = H'H$. We discuss this point elsewhere.

Dynamic Programing Formulation

With A, B, Q and R constant, the algebraic Riccati eqaution can also be derived by application of the principle of optimality as we next demonstrate. We know that $x_0'Px_0 = \int_0^\infty (x_t'Qx_t + u_t'Ru_t)dt$. Break up the integral into two parts; one over $[0, \Delta)$ and the other $[\Delta, \infty)$ so that

$$\text{RHS} = (x_0'Qx_0 + u_0'Ru_0)\Delta + \int_\Delta^\infty (x_t'Qx_t + u_t'Ru_t)dt.$$

Because the problem is time-invariant, we can write the latter as

$$\int_\Delta^\infty (x_t'Qx_t + u_t'Ru_t)dt = x_\Delta'Px_\Delta.$$

The principle of optimality states that

$$x_0'Px_0 = \operatorname*{Min}_{u_0}\{(x_0'Qx_0 + u_0'Ru_0)\Delta + x_\Delta'Px_\Delta + o(\Delta)\}$$

where

$$x_\Delta' = x_0 + (Ax_0 + Bu_0)\Delta + o(\Delta)$$

or

$$x_0'Px_0 = \operatorname*{Min}_{u_0}\{(x_0'Qx_0 + u_0'Ru_0)\Delta + x_0'Px_0$$

$$+ (Ax_0 + Bu_0)'Px_0\Delta + x_0'P(Ax_0 + Bu_0)\Delta + o(\Delta)\}.$$

Cancelling the term $x_0'Px_0$ from both side, dividing by Δ and letting it

Approach zero, we obtain

$$0 = \min_{u_0} \{x_0'Qx_0 + u_0'Ru_0 + (Ax_0 + bu_0)'Px_0 + x_0'P(Ax_0 + Bu_0)\}.$$

On carrying out the minimization, we discover that

$$u_0 = -R^{-1}B'Px_0.$$

Because the problem is time-invariant, we know that in general

$$u_t = -R^{-1}B'Px_t.$$

Substituting the optimal control yeilds the same matrix Riccati equation for P given by (8).

A.16 Maximum Principle: Discrete-Time Dynamics

A typical problem formulation is to minimize a sum of nonlinear functions

$$J_T = \Sigma_0^{T-1} f_k^0(z_k, x_k),$$

subject to dynamic constraints

$$z_{k+1} = z_k + f_k(z_k, x_k), \qquad k = 0, 1, \ldots, T-1,$$

where $x_k \in U_k \subset E^r$.

The set U_k is the set from which decision vectors are to be selected. Here we assume that $f_0, \ldots, f_{T-1}$ are continuously differentiable in both arguments. In addition, further constraints may be imposed on the state vectors to satisfy some inequality and equality equations:

$$z_0 \in Z_0 = Z_0' \cap Z_0''$$

where

$$Z_0' = \{z: \ q_0(z) \leq 0\},$$

$$Z_0'' = \{z: \ g_0(z) = 0\},$$

$$z_t \in Z_t' = \{z: \ q_t(z) \leq 0\}$$

and

$$z_T \in Z_T = Z_T' \cap Z_T'',$$

where

$$Z_T' = \{z: \ q_T(z) \leq 0\},$$

$$Z_T'' = \{z: \ g_T(z) = 0\}.$$

We assume that, unless identically zero, $\partial g_0/\partial z$ and $\partial g_T/\partial z$ have maximum ranks on Z_0 and Z_T, respectively. Additionally, technical conditions are needed to ensure that gradient vectors of active constraints are linearly independent (so that Z_t', $t = 0, \ldots, T$ may have a right kind of conical approximations) (see Canon et al. [1970]), and to have conical approximations for the set U_t, $t = 0, 1, \ldots, T-1$.

With these assumptions we can state the necessary condition for optimal decisions which is the discrete dynamics version of the Pontryagin's maximum principle. (Canon et al. [1970])

Theorem (Discrete-Maximum Principle)

There exist costate vectors, $p_0, p_1, \ldots, p_{T-1}$ in E^n, multiplier vectors $\lambda_0, \lambda_1, \ldots, \lambda_T, \lambda_t \leq 0, \mu_0, \mu_T$, and a scalar $p^0 \leq 0$, such that (i) not all $p^0, p_0, p_1, \ldots, p_T$ and μ_0, μ_T are zero,

(ii) p_t satisfies

$$p_t - p_{t+1} = [\partial f_t(z_t^*, x_t)/\partial z]' p_{t+1} + p^0 [\partial f_t^0(z_t^*, x_t/\partial z]' + [\partial q_t(z_t^*)/\partial z]' \lambda_t, \qquad t = 0, 1, \ldots, T-1,$$

(iii) subject to the transversality conditions

$$p_0 = - [\partial g_0(z_0)/\partial z]' \mu_0$$

$$p_T = [\partial g_T(z_T)/\partial z]' \mu_0 + [\partial q_T(z_T)/\partial z]' \lambda_T$$

and

$$(\lambda_t, q_t(z_t^*)) = 0, \qquad t = 0, 1, \ldots, T$$

and

(iv) x_t maximizes $H(z, x, p, p^0, t) = p^0 f_t^0(z, x) + \langle p, f_t(z, x) \rangle$ i.e.,

$$H(z_t^*, x_t^*, p_{t+1}, p^0, t) \geq H(z_t^*, x_t, p_{t+1}, p^0, t) \text{ for all } x_t \in U_t.$$

A.17 Policy Reaction Functions, Stabilization Policy and Modes

Lucas' critique (Lucas [1976]) of a common econometric practice seems to have had much impact on the econometric profession. For those who have known that (state variable) feedbacks modify dynamic characteristics his critique is not a bit surprising. In fact policy reaction functions or any sort of feedbacks between policies and the "state" of the economy will modify the locations of the eigenvalues which govern (linearized or variational) dynamics of the economy, hence change the structure of dynamics.

For a simple illustration of this point, let

$$z = Az + bu$$

where b is an n-dimensional vectors, and u is a scalar exogenous variable. The dynamic characteristcs are determined by the eigenvalues of A which are the roots of the characteristic equation $a(\lambda) = |\lambda I - A| = \lambda^n + a_1\lambda^{n-1} + \dots + a_n$. Now suppose that u is generated by $k'z + v$ where $k'z$ is an automatic, i.e., a reaction part of the policy variable and v is the discretionary or exogenous part. The differential equation now changes into

$$\dot{z} = (A + bk')z + bv.$$

The eigenvalues of $A + bk'$ are the roots of $d_k(\lambda) = |\lambda I - A - bk'|$. Rewrite it as manipulated as follows (see Bass and Gura [1965]) to reveal the effects of k more clearly:

$$(+)\qquad \begin{aligned} d_k(\lambda) &= |\lambda I - A||I - (\lambda I - A)^{-1}bk'| \\ &= d(\lambda)\{1 + k'(\ I - A)^{-1}b\}. \end{aligned}$$

The second line follows from a well-known matrix identify that $|I + ab'| = 1 + b'a$. (See Aoki [1976; p.389], for example.)

Next, note that $(\lambda I - A)^{-1}$ can be written as

$$(\lambda I - A)^{-1} = \text{adj}(\lambda I - A)/d(\lambda)$$

where the matrices $B_1, \ldots, B_{n-1}$ in the expression

$$\text{adj}(\lambda I - A) = \lambda^{n-1}I + B_1\lambda^{n-2} + \ldots + B_{n-1}$$

are recursively generated by equating the like powers in λ*

$$d(\lambda)I = (\lambda I - A)\text{adj}(\lambda I - A),$$

i.e.,

$$B_i = B_{i-1}A + a_{i-1}I \qquad i = 1, \ldots, n-1$$

$$B_0 = I$$

or $$B_1 = A + a_1 I,\ B_2 = A^2 + a_1 A + a_2 I \quad \text{etc.}$$

where

$$d(\lambda) = \lambda^n + a_1\lambda^{n-1} + \ldots + a_n.$$

Equation (+) now states that

$$d_k(\lambda) - a(\lambda) = (k'b)\lambda^{n-1} + k'(A + a_1 I)b\lambda^{n-2} + \ldots + k'B_{n-1}b.$$

Let $a_k(\lambda)$ be $\lambda^n + \beta_1\lambda^{n-1} + \ldots + \beta_n$. Then

$$\beta_i - a_i = k'B_i b, \qquad i = 1, \ldots, n$$

when B_i is are substituted out, we can arrange them

$$(\beta_1, \ldots, \beta_n) - (a_1, \ldots, a_n) = k[b, Ab, \ldots, A^{n-1}b]\begin{bmatrix} 1 & a_1 & \cdots & a_{n-1} \\ & \ddots & \ddots & \vdots \\ & & \ddots & a_1 \\ 0 & & & 1 \end{bmatrix}.$$

If a desired configuration of eigenvalues are specified, then β's are determined. If the matrix $[b, Ab, \ldots, A^{n-1}b]$ is nonsingular, then the above equation can uniquely be solved for k given β_i's and a_i's, i.e., the desired eigenvalue pattern are realized with this k. Or, the equation can be read to state that for a given choice of k, the dynamics of A is modified in such a way that the coefficients of the characteristic polynomials satify the above equation.

---

* This is known as Leverrier's algorithm (Aoki [1976; p.45]).

Dynamic characteristcs are then influenced by policy regimes. As Sargent states, the matrix A should not have been regarded as "structural" parameters since they do not remain invariant in the face of policy interventions.

A.18 Dynamic Policy Multipliers

Effects of changing instruments on endogenous variables figure importantly in any contemplated policy changes. Not only the contemporaneous and long-run effects i.e., the effects felt at the same time of instrument changes and when all adjustments have worked themselves out in the model, but the processes of adjustment must be evaluated. Dynamic multiplier do precisely that. Elasticity is intended to tell us how the economy will respond to a change in a given situation. Naturally this response will depend on the situation in general. Variational dynamics we later introduce generalize the notion of elasticity to dynamic context, because variational dynamics tell us how the economy will respond over time to changes in a given situation, where the situation is to be understood dynamically, i.e., the situation is a set of time paths of endogenous and instrument variables. We have called them reference time paths. Just as elasticity depends on the situation, time behavior of variational dynamics will depends on the time paths of the situation. Dynamic multipliers depict the time profile of policy effects, and can be thought as dynamic elasticity concept to dynamic situations. Dynamic multipliers, therefore, is a particular example of elasticity of a time path.

To appreciate simplicity of state space representaion, try calculating dynamic multipliers using the two alternate representaions i.e., calculate effects of changing inputs at time τ by some finite amout on the contemporaneous outputs, i.e., outputs at time τ as well as on all subsequent times, $t > \tau$, using models in state space form and in ARMA form. This comparison should convince the reader of messier manipulation needed by the ARMA representation. (Aoki [1981; Appendix A] conducts such a comparison.) Also intertemporal optimization is easier to conduct when dynamic models are in state space form, especially using dynamic programming.

Dynamic multiplier calculations on state space models typically proceed as follows: We wish to compare effects of two alternate input sequences $\{u_t^i\}$, i = 1, 2 on y_t using a model

$$\begin{cases} x_{t+1}^i = Ax_t^i + Bu_t^i, \\ y_t^i = Cx_t^i + Du_t^i \end{cases} \qquad i = 1, 2.$$

Then

$$y_t^1 - y_t^2 = C(x_t^1 - x_t^2) + D(u_2^1 - u_t^2),$$

where

$$x_t^1 - x_t^2 = \Sigma_{t=1}^{t-1} A^{t-1-\tau} B(u_\tau^1 - u_\tau^2),$$

because, to be fair, we must compare two alternative policies on the system starting from the same state at the initial time 0. Suppose

$$u_\tau^1 - u_\tau^2 = \{\begin{matrix} 0 & \tau \leq s, & \tau > s \\ \Delta u & \tau = s & \end{matrix} \quad .$$

Then

$$y_s^1 - y_s^2 = D\Delta u,$$

and

$$y_t^1 - y_t^2 = CA^{t-1-s}B\Delta u \qquad t > s.$$

This expression $CA^{t-1-s}B$ is the dynamic multiplier (matrix). It is also known as the Markovian parameters. We later return to this expression, because they are also important in constructing state space models to (approximately) reproduce given input and output data sequences.

REFERENCES

Achieser, N. I. (1956), Theory of Approximation, translated by C. J. Hyman, F. Ungar Pub. Co., New York.

Akaike, H. (1973), "Maximum Likelihood Identification of Gaussian Autoregressive Moving Average Models," Biometrika, 60, 255-265.

__________ (1974), "Markovian Representation of Stochastic Processes and Its Application to the Analysis of Autoregressive Moving Average Processes," Ann. Inst. Statist. Math., 26, 363-387.

__________ (1976), "Canonical Correlation Analysis of Time Series and the Use of an Information Criterion," in, R. Mehra and D. Lainiotis, eds., System Identification: Advances in Case Studies, Academic Press, Inc., New York.

__________ (1980), "Seasonal Adjustment by a Bayesian Modeling," J. Time Series Analysis, 1, 1-13.

Allen, R. G. D. (1966), Mathematical Economics, 2nd ed., Macmillan, London.

Anderson, B. D. O. and S. Vongparitlerd (1973), Network Analysis and Synthesis, Prentice-Hall, New Jersey.

__________, K. L. Hitz and N. D. Dieu (1974), "Recursive Algorithm for Spectral Factorization," IEEE Trans. Circuits and Systems, CAS-21, 742-750.

__________ and J. B. Moore (1975), Optimal Filtering, Prentice-Hall, New Jersey.

Aoki, M. (1962), "On a Successive Approximation Technique in Solving Some Control System Optimization Problems," J. Math. Anal. Appl., 15, 418-434.

__________ (1964), "On Optimal and Suboptimal Control Policies in Control Systems" in, C.T. Leondes (ed.) Advances in Control Systems 1, Academic Press, Inc., New York.

__________ (1967), Optimization of Stochastic Systems, Academic Press, Inc., New York.

__________ (1968), "Control of Large Scale dynamic Systems by Aggregation," IEEE Trans. Aut. Control, AC-13, 246-253.

__________ (1968), "Note on Aggregation and Bounds for Solution of the Matrix Riccati Equation," J. Math. Anal. Appl., 21, 379-383.

__________ and R. M. Staley (1970), "On Input Signal Synthesis in Parameter Identification," Automatica, 6, 431-440.

__________ and P. C. Yue (1970), "On a Priori Error Estimates of Some Identification Methods," IEEE Trans. Aut. Control, AC-15, 541-548.

__________ (1971), Introduction to Optimization Techniques; Fundamentals and Applications of Nonlinear Programing, Macmillan, New York.

__________ (1973), "On subspaces Associated with Partial Reconstruction of State Vectors, the Structure Algorithm, and the Pardictable Directions of Ricatti Equations," IEEE, AC-18, 399-400.

M. Aoki and M. T. Li (1973), "Partial Reconstruction of State Vectors in Decentralized Dynamic Systems," IEEE, AC-18, 289-292.

__________ (1976), Optimal Control and System Theory in Dynamic Analysis, North-Holland, Amsterdam.

__________ and M. Canzoneri (1979), "Reduced Forms of Rational Expectations Models," QJE, 93, 59-71.

__________ (1980), "Comparative Dynamic Analysis of a Growth Model under Alternative Policy Regimes," J. Macroeconomics, 2, 1-39.

__________ (1981), Dynamic Analysis of Open Economics, Academic Press, Inc., New York.

Bass, R. W. and I. Gura (1965), "High Order Design via State-Space Considerations," Proceedings of 1965 JACC, Troy, New York, 311-318.

Baumol, W. J. (1970), Economic Theory and Operations Analysis, 3rd ed., Prentice-Hall, New Jersey.

Bellman, R.E. (1957), Dynamic Programming, Princeton University Press, New Jersey.

__________ (1960), Introduction to Matrix Analysis, McGraw-Hill, New York.

__________ and S. E. Dreyfus (1962), Applied Dynamic Programming, Princeton University Press, New Jersey.

Birkhoff, G. and S. MacLane (1970), A Survey of Modern Algebra, Macmillan, New York.

Blanchard, O. J. (1979), "Backward and Forward Solution for Economics with Rational Expectations," Am. Econ. Rev., 69, 114-118.

Blinder, A. S. and S. Fisher (1981), "Inventories, Rational Expectations and in the Business Cycle," J.M.E., 8, 277-304.

Bosgra, O.H., A. J. J. Van der Weiden (1980), "Innut-Output Invariants for Linear Multivariable Systems," IEEE Trans., AC-25, 20-36.

Box, G. E. and G. M. Jenkins (1970), Time-Series Analysis, Forecasting and Control, Holden-Day, San Francisco.

Bryson, A. E. and L. J. Henrikson (1968), "Estimation Using Sampled Data Containing Sequentially Correlated Noise," J. Spacecraft and Rockets, 5, 662-666.

Canon, M. D., C. D. Culum, Jr., and E. Polak (1970), Theory of Optimal Control and Mathematical Programming, McGraw-Hill, New York.

Canzoneri, M. B. and J. M. Underwood (1983), "Wage Contractings, Exchange Rate Volatility, and Exchange Interventions Policy," to be presented at 1983 annual meeting of Soc. Econ. Dynamics and Control.

Chen, C. T. (1970), Analysis and Synthesis of Linear Control Systems, Holt, Reinhard & Winston, New York, Chapts. 2, 4, and 8.

Chow, G. C. (1975), Analysis and Control of Dynamic Economic Systems, John Wiley, New York.

Davenport, W. B. Jr., and W. L. Root (1958), An Introduction to the Theory of Random Signals and Noise, McGraw-Hill, New York.

Deistler, M., W. Pliberger and B. M. Rötscher (1982), "Identifiability and Inference in ARMA Systems," in Time Series Analysis: Theory and Practice 2, O. D. Anderson, ed., North-Holland, Amsterdam.

Denham, M. J. (1974), "Canonical Forms for the Identification of Multivariable Systems," IEEE Trans., AC-19, 646-656.

Dunsmuir, W. and E. J. Hannan (1976), "Vector Linear Time Series Models," Adv. Appl. Prob., 8, 339-364.

Faurre, P.L. (1976), "Stochastic Realization Algorithms" in, R. Mehra and D. Lainiotis, eds., opt. cit.

Fleming, W. H. and R. W. Rishel (1975), Deterministic and Stochastic Optimal Control, Springer-Verlag, Berlin.

Futia, C. A. (1979), "Stochastic Business Cycles," Bell Telephone Labs., Tech. Report.

Gandolfo, G. (1971), Mathematical Methods and Models in Economic Dynamics, North-Holland, Amsterdam.

Geyers, M. R. and T. Kailath (1973), "An Innovation Approach to Least-Squares Estimation, Part VI: Discrete Time Innovation Representation and Recursive Estimation," IEEE, AC-18, 588-600.

Glover, K. and J. C. Willems (1974), "Parametrizations of Linear Dynamical Systems: Canonical Forms and Identifiability," IEEE, AC-19, 640-646.

Golub, G. H. and C. Reinsch (1970), "Singular Value Decomposition and Least Squares Solutions," Numer. Mat., 14, 403-420.

Goodwin, G. C. and R. L. Payne (1977), Dynamic System Identification: Experiment Design and Data Analysis, Academic Press, Inc., New York.

Gourieroux, C., J. J. Laffont and A. Monfort (1979), "Rational Expectations Models: Analysis of the Solutions," INSEE Mimeo, May.

Granger, C. W. J. and P. Newbold (1977), Forecasting Economic Time Series, Academic Press, Inc., New York.

Guidorzi, R. P. (1981), "Invariants and Canonical Forms for Systems Structural and Parametric Identification," Automatica, 17, 117-133.

Hall, R. E. (1978), "Stochastic Implications of the Life Cycle-Permanent Income Hypothesis," J.P.E., 6, 971-988.

Hannan, E. J. and J. Rissanen (1982), "Recursive Estimation of Mixed Autoregressive Moving Average Order," Biometrika, 69, 81-94.

Hansen, L. P. and T. J. Sargent (1980), "Formulating and Estimating Dynamic Rational Expectations Models," J. Econ. Dynamics and Control, 2, 7-46.

__________ and K. J. Singleton (1982), "Generalized Instrumental Variables Estimation of Nonlinear Rational Expectations Models," Econometrica, 50, 1269-1286.

Harvey, A. C. and G. Stein (1978), "Quadratic Weights for Asymptotic Regulator Properties," IEEE, AC-23, 378-387.

__________ (1981), The Econometric Analysis of Time Series, Philip Allan Pub., Oxford.

Hewer, G. A. (1971), "An Iterative Technique for the Computation of the Steady State Gains for the Discrete Optimal Regulator," IEEE, AC-16, 382-383.

Hodrick, R. J. and E. C. Prescott (1981), "Post-war U.S. Business Cycles: An Empirical Investigation," Discussion Paper No.451, Carnegie-Mellon University.

Ito, T. (1982), "A Comparison of Japanese and U.S. Macroeconomic Behavior by a VAR Model," Discussion Paper No. 82-162, University of Minnesota.

Kailath, T. (1980), Linear System Theory, Prentice-Hall, New Jersey.

Kamien, M. F. and Schwartz, N. L. (1981), Dynamic Optimization, North-Holland, Amsterdam.

Kimura, H. (1982), Digital Signal Processing and Control (in Japanese), Shokodo, Tokyo.

Kshirsagar, A. M. (1972), Multivariate Analysis, Marcel Dekker, New York.

Kucera, V. (1972), "The Discrete Riccati Equation of Optimal Control," Kybernetika (Prague), 8, 430-447.

Kung, S-Y. and D. W. Lin (1981), "Optimal Hankel-Norm Model Reductions: Multivariable Systems," IEEE Trans. Aut. Control, AC-26, 832-852.

Kwakernaak, H. and R. Sivan (1972), Linear Optimal Control Systems, John Wiley, New York.

Kydland, F. and E. C. Prescott (1981), "Time to Build and Aggregate Fluctuations," Carnegie Mellon Working Paper, 28-80-81.

Laning, J. H., Jr. and R. H. Battin (1956), Random Processes in Automatic Control, McGraw Hill, New York.

Lee, E. P. and L. Markus (1967), Founations of Optimal Control Theory, John Wiley, New York.

Long, J.B., Jr. and C. I. Plosser (1983), "Real Business Cycles," J.P.E., 91, 39-69.

Lucas, R. E., Jr. (1975), "An Equilibrium Model of the Business Cycle," J.P.E., 83, 1113-1144.

__________ (1976), "Econometric Policy Evaluation: A Critique" in K. Brunner and A. H. Metlzer, eds., The Phillips Curve and Labor Markets, Carnegie-Rochester Conferences on Public Policy Vol., North-Holland, Amsterdam.

__________ (1977), "Understanding Business Cycles," in Carnegie-Rochester Conference Series on Public Policy, 5, Stabilization of the Domestic and International Economy, 7-29.

Luenberger, D.G. (1979), Introduction to Dynamic Systems, John Wiley, New York.

MacFarlane, A. G. J. (1979), "The Development of Frequency-response Methods in Automatic Control," IEEE, AC-24, 250-265.

Molinari, B. P. (1975), "The Stabilizing Solution of the Discrete Algebraic Riccati e.q.," IEEE Trans., AC-20, 396-399.

Moore, B. C. (1976), "On the Flexibility Offered by State Feedback in Multivariable Systems Beyond Closed Loop Eigenvalue Assignment," IEEE, AC-21, 689-692.

__________ (1978), "Singular Value Analysis of Linear Systems, Part I and II," Proc. 1978 IEEE Conference on Decision and Control, 66-73.

Moore, J. B. and B. D. O. Anderson (1968), "Extensions of Quadratic Minimization Theory," I: Finite Time Results, II: Infinite Time Results, International J. Control, 7, 465-480.

Newton, G. C., L. A. Gould and J. F. Kaiser (1957), Analytical Design of Linear Feedback Controls, John Wiley, New York.

Pagano, M. (1976), "On the Linear Convergence of a Covariance Factorization Algorithm," J. Ass. Comp. Math., 23, 310-316.

Pernebo, L. and L. M. Silverman (1982), "Model Reduction via Balanced State Representations," IEEE, AC-27, 382-387.

Picci, G. (1982), "Some Numerical Aspects of Multivariable Systems Identification," Math. Progr. Study, 18, 76-101.

Priel, B. and U. Shaked (1983), "Cheap Optimal Control of Discrete Single Input Single Output Systems," unpublished mimeo, Tel-Aviv University.

Rao, C. Radhaknishna (1964), "The Use and Interpretation of Principal Component Analysis in Applied Research," Sankya Series A, 26, 329-358.

Rissanen, J. (1974), "Basis of Invariants and Canonical Forms for Linear Dynamical Systems," Automatica, 10, 175-182.

__________ (1976), "Minimax Entropy Estimation of Models for Vector Process," in R. Mehra and D. Lainiotis, eds., opt. cit.

Robinson, E. A. and S. Treitel (1980), Geophsical Signal Analysis, Prentice-Hall, New Jersey.

Rosenbrock, H. H. (1970), State-Space and Multivariable Theory, Thomas Nelson and Sons Ltd., London.

Sargent, T. J. (1979), Macroeconomic Theory, Academic Press, Inc., New York.

__________ (1981), "Interpreting Economic Time Series," Journal of Political Economy, 89, 213-248.

Schwarz, G. (1978), "Estimating the Dimension of a Model," Annals of Statistics, 6, 461-464.

Shaked, U. (1979), "A Transfer Function Approach to the Linear Discrete Stationary Filtering and Steady State Optimal Control Problems," International Journal of Control, 29, 279-291.

Shiskin, J. and T. J. Plewes (1978), "Seasonal Adjustment of the U.S. Unemployment Rate," Statistician, 27, 181-202.

Sims, C. A. (1980), "Macroeconomics in Reality," Econometrica, 48, 1-48.

____________ (1980), "Comparison of Interwar and Postwar Business Cycles: Monetarism Reconsidered," AER, 70, 250-257.

____________ (1982), "Policy Analysis with Econometric Models," Brookings Papers on Economic Activity, 107-152.

Solo, Victor (1983), Topics in Advanced Time Series Analysis, Lecture Notes, Springer-Verlag, New York.

Son, L. H. and B. D. O. Anderson (1973), "Design of Kalman Filters Using Signal Model Output Statistics," Proc. IEE, 120, 312-318.

Stein, G. (1979), "Generalized Quadratic Weights for Asymptotic Regulator Properties," IEEE, AC-24, 559-566.

Strang, G. (1973), Linear Algebra and its Applications, Academic Press, Inc., New York.

Takeuchi, K. (1983), "On Statistical Model Selection based on AIC," (in Japanese) J. Soc. Instrument & Control Engineers, 22, 445-453.

Van Zee, G. A. (1981), System Identification for Multivariable Control, Delft University Press.

Wertz, V. (1981), "Structure Selection for the Identification of Multivariate Process," Catholic University of Leuven Ph. D. Thesis in Applied Science.

Whittle, P. (1963), Prediction and Regulation by Linear Least-Square Methods, Van Nostrand, Princeton, New Jersey.

____________ (1982), Optimization Over Time, Vol.1, John Wiley, New York.

Wonham, W. M. (1967), "On Pole Assignment in Multi-Input Controllability Linear Systems," IEEE, AC-12, 660-665.

INDEX

Vol. 157: Optimization and Operations Research. Proceedings 1977. Edited by R. Henn, B. Korte, and W. Oettli. VI, 270 pages. 1978.

Vol. 158: L. J. Cherene, Set Valued Dynamical Systems and Economic Flow. VIII, 83 pages. 1978.

Vol. 159: Some Aspects of the Foundations of General Equilibrium Theory: The Posthumous Papers of Peter J. Kalman. Edited by J. Green. VI, 167 pages. 1978.

Vol. 160: Integer Programming and Related Areas. A Classified Bibliography. Edited by D. Hausmann. XIV, 314 pages. 1978.

Vol. 161: M. J. Beckmann, Rank in Organizations. VIII, 164 pages. 1978.

Vol. 162: Recent Developments in Variable Structure Systems, Economics and Biology. Proceedings 1977. Edited by R. R. Mohler and A. Ruberti. VI, 326 pages. 1978.

Vol. 163: G. Fandel, Optimale Entscheidungen in Organisationen. VI, 143 Seiten. 1979.

Vol. 164: C. L. Hwang and A. S. M. Masud, Multiple Objective Decision Making – Methods and Applications. A State-of-the-Art Survey. XII, 351 pages. 1979.

Vol. 165: A. Maravall, Identification in Dynamic Shock-Error Models. VIII, 158 pages. 1979.

Vol. 166: R. Cuninghame-Green, Minimax Algebra. XI, 258 pages. 1979.

Vol. 167: M. Faber, Introduction to Modern Austrian Capital Theory. X, 196 pages. 1979.

Vol. 168: Convex Analysis and Mathematical Economics. Proceedings 1978. Edited by J. Kriens. V, 136 pages. 1979.

Vol. 169: A. Rapoport et al., Coalition Formation by Sophisticated Players. VII, 170 pages. 1979.

Vol. 170: A. E. Roth, Axiomatic Models of Bargaining. V, 121 pages. 1979.

Vol. 171: G. F. Newell, Approximate Behavior of Tandem Queues. XI, 410 pages. 1979.

Vol. 172: K. Neumann and U. Steinhardt, GERT Networks and the Time-Oriented Evaluation of Projects. 268 pages. 1979.

Vol. 173: S. Erlander, Optimal Spatial Interaction and the Gravity Model. VII, 107 pages. 1980.

Vol. 174: Extremal Methods and Systems Analysis. Edited by A. V. Fiacco and K. O. Kortanek. XI, 545 pages. 1980.

Vol. 175: S. K. Srinivasan and R. Subramanian, Probabilistic Analysis of Redundant Systems. VII, 356 pages. 1980.

Vol. 176: R. Färe, Laws of Diminishing Returns. VIII, 97 pages. 1980.

Vol. 177: Multiple Criteria Decision Making-Theory and Application. Proceedings, 1979. Edited by G. Fandel and T. Gal. XVI, 570 pages. 1980.

Vol. 178: M. N. Bhattacharyya, Comparison of Box-Jenkins and Bonn Monetary Model Prediction Performance. VII, 146 pages. 1980.

Vol. 179: Recent Results in Stochastic Programming. Proceedings, 1979. Edited by P. Kall and A. Prékopa. IX, 237 pages. 1980.

Vol. 180: J. F. Brotchie, J. W. Dickey and R. Sharpe, TOPAZ – General Planning Technique and its Applications at the Regional, Urban, and Facility Planning Levels. VII, 356 pages. 1980.

Vol. 181: H. D. Sherali and C. M. Shetty, Optimization with Disjunctive Constraints. VIII, 156 pages. 1980.

Vol. 182: J. Wolters, Stochastic Dynamic Properties of Linear Econometric Models. VIII, 154 pages. 1980.

Vol. 183: K. Schittkowski, Nonlinear Programming Codes. VIII, 242 pages. 1980.

Vol. 184: R. E. Burkard and U. Derigs, Assignment and Matching Problems: Solution Methods with FORTRAN-Programs. VIII, 148 pages. 1980.

Vol. 185: C. C. von Weizsäcker, Barriers to Entry. VI, 220 pages. 1980.

Vol. 186: Ch.-L. Hwang and K. Yoon, Multiple Attribute Decision Making – Methods and Applications. A State-of-the-Art-Survey. XI, 259 pages. 1981.

Vol. 187: W. Hock, K. Schittkowski, Test Examples for Nonlinear Programming Codes. V. 178 pages. 1981.

Vol. 188: D. Bös, Economic Theory of Public Enterprise. VII, 142 pages. 1981.

Vol. 189: A. P. Lüthi, Messung wirtschaftlicher Ungleichheit. IX, 287 pages. 1981.

Vol. 190: J. N. Morse, Organizations: Multiple Agents with Multiple Criteria. Proceedings, 1980. VI, 509 pages. 1981.

Vol. 191: H. R. Sneessens, Theory and Estimation of Macroeconomic Rationing Models. VII, 138 pages. 1981.

Vol. 192: H. J. Bierens: Robust Methods and Asymptotic Theory in Nonlinear Econometrics. IX, 198 pages. 1981.

Vol. 193: J. K. Sengupta, Optimal Decisions under Uncertainty. VII, 156 pages. 1981.

Vol. 194: R. W. Shephard, Cost and Production Functions. XI, 104 pages. 1981.

Vol. 195: H. W. Ursprung, Die elementare Katastrophentheorie. Eine Darstellung aus der Sicht der Ökonomie. VII, 332 pages. 1982.

Vol. 196: M. Nermuth, Information Structures in Economics. VIII, 236 pages. 1982.

Vol. 197: Integer Programming and Related Areas. A Classified Bibliography. 1978 – 1981. Edited by R. von Randow. XIV, 338 pages. 1982.

Vol. 198: P. Zweifel, Ein ökonomisches Modell des Arztverhaltens. XIX, 392 Seiten. 1982.

Vol. 199: Evaluating Mathematical Programming Techniques. Proceedings, 1981. Edited by J.M. Mulvey. XI, 379 pages. 1982.

Vol. 201: P. M. C. de Boer, Price Effects in Input-Output-Relations: A Theoretical and Empirical Study for the Netherlands 1949–1967. X, 140 pages. 1982.

Vol. 202: U. Witt, J. Perske, SMS – A Program Package for Simulation and Gaming of Stochastic Market Processes and Learning Behavior. VII, 266 pages. 1982.

Vol. 203: Compilation of Input-Output Tables. Proceedings, 1981. Edited by J. V. Skolka. VII, 307 pages. 1982.

Vol. 204: K.C. Mosler, Entscheidungsregeln bei Risiko: Multivariate stochastische Dominanz. VII, 172 Seiten. 1982.

Vol. 205: R. Ramanathan, Introduction to the Theory of Economic Growth. IX, 347 pages. 1982.

Vol. 206: M.H. Karwan, V. Lotfi, J. Telgen, and S. Zionts, Redundancy in Mathematical Programming. VII, 286 pages. 1983.

Vol. 207: Y. Fujimori, Modern Analysis of Value Theory. X, 165 pages. 1982.

Vol. 208: Econometric Decision Models. Proceedings, 1981. Edited by J. Gruber. VI, 364 pages. 1983.

Vol. 209: Essays and Surveys on Multiple Criteria Decision Making. Proceedings, 1982. Edited by P. Hansen. VII, 441 pages. 1983.

Vol. 210: Technology, Organization and Economic Structure. Edited by R. Sato and M.J. Beckmann. VIII, 195 pages. 1983.

Vol. 211: P. van den Heuvel, The Stability of a Macroeconomic System with Quantity Constraints. VII, 169 pages. 1983.

Vol. 212: R. Sato and T. Nôno, Invariance Principles and the Structure of Technology. V, 94 pages. 1983.

Vol. 213: Aspiration Levels in Bargaining and Economic Decision Making. Proceedings, 1982. Edited by R. Tietz. VIII, 406 pages. 1983.

Vol. 214: M. Faber, H. Niemes und G. Stephan, Entropie, Umweltschutz und Rohstoffverbrauch. IX, 181 Seiten. 1983.

Vol. 215: Semi-Infinite Programming and Applications. Proceedings, 1981. Edited by A. V. Fiacco and K. O. Kortanek. XI, 322 pages. 1983.

Vol. 216: H. H. Müller, Fiscal Policies in a General Equilibrium Model with Persistent Unemployment. VI, 92 pages. 1983.

Vol. 217: Ch. Grootaert, The Relation Between Final Demand and Income Distribution. XIV, 105 pages. 1983.

Vol. 218: P. van Loon, A Dynamic Theory of the Firm: Production, Finance and Investment. VII, 191 pages. 1983.

Vol. 219: E. van Damme, Refinements of the Nash Equilibrium Concept. VI, 151 pages. 1983.

Vol. 220: M. Aoki, Notes on Economic Time Series Analysis: System Theoretic Perspectives. IX, 249 pages. 1983.